DISCARDED

NON-CIRCULATING MATERIAL

NON-CIRCULATING MATERIAL

Advances in Fluidization and Fluid Particle Systems

Desmond King, Volume Editor
Hamid Arastoopour, John C. Chen, Alan W. Weimer and Wen-Ching Yang, Volume Co-Editors

S. Alsop	J.R. Grace	S. Krol	A. Sahnoun
M. Aufauvre	J.S. Groen	J.F. Large	Jack Saluja
H. Arastoopour	P. Guigon	Lii-ping Leu	J.C. Schouten
J. Baxter	Bernd Hage	K.S. Lim	S. Shao
J.J. L. Berben	D.M. Heyes	William Link	C.C. Shie
F. Berruti	T.C. Ho	H. Littman	D. Smith
C.M. van den Bleek	J.R. Hoppes	Shi Liu	C.S. Stellema
Robert C. Brown	Y.K. Hsia	X. Loison	E.A. Stephan
Ethan Brue	Lu Huilin	Antonio Marzocchella	J. Su
G.G. Chase	A.S. Issangya	David J. Mason	Stephen Tallon
C.C. Chen	Heinrich M. Jaeger	A.J. Matchett	Jonathan Thorn
George D. Cody	Hongshen Ji	C.B. Morgan	Atsushi Tsutsumi
Charles J. Coronella	Filip Johnsson	M.H. Morgan, III	U. Tüzün
J.M. Coulthard	N.K. Kimm	R.F. Mudde	H.E.A. Van Den Akker
Clive E. Davies	George Klinzing	Sidney R. Nagel	C.M. van den Bleek
H. de Lasa	James B. Knight	Volen R. Nikolov	R. Vengimalla
Jianxun Deng	Ted M. Knowlton	Iordan Nikov	K.S. Wang
Ihab H. Farag	Z.I. Kolar	R. Ocone	Herbert Weinstein
M.A. Forcinito	Alain Koniuta	G. Papavasiliou	Joachim Werther
J.J.M. de Goeij	H.O. Kono	A. Pekediz	M.S. Willis
T.C. Ho	John Kost	D. Rai	V.K.N. Yerra
Dimitri Gidaspow	A.E. Kostazos	L. Richman	Kunio Yoshida
David J. Goldfarb			Robert C. Zijerveld

AIChE Staff
Maura Mullen, Managing Editor
Cover Design: Armand Veneziano

AIChE Symposium Series

Number 317   1997   Volume 93

Published by

American Institute of Chemical Engineers

345 East 47 Street   New York, N.Y. 10017

© 1997
American Institute of Chemical Engineers (AIChE)
345 E 47 Street
New York. N.Y. 10017

AIChE shall not be responsible for statements or opinions advanced in their papers or printed in their publications.

ISBN 0-8169-0742-0

All rights reserved whether the whole or part of the material is concerned, specifically those of translation, reprinting, re-use of illustrations, broadcasting, electronic networks, reproduction by photocopying machine or similar means, and storage of data in banks.

Authorization to photocopy items for internal use, or the internal or personal use of specific clients, is granted by AIChE for libraries and other users registered with the Copyright Clearance Center Inc., provided that the $3.00 base fee + $1.50 per page is paid directly to CCC, 222 Rosewood Drive, Danvers, MA 01923. This consent does not extend to copying for general distribution, for advertising, or promotional purposes, for inclusion in a publication, or for resale.

Articles published before 1978 are subject to the same copyright conditions and the fee is $3.50 for each article. AIChE Symposium Series fee code: 0065-8812/1997.

FOREWORD

This volume of the AIChE Symposium Series continues the tradition of an annual volume presenting recent developments in fluidization and fluid-particle systems. The papers are selected by peer reviews from those presented in ten sessions at the AIChE Annual Meeting in Chicago, Illinois, November 10 through 15, 1996.

This year's volume includes papers that cover a wide range of topics related to fluid-particle systems:

- Fluidization Fundamentals
- Circulating Fluidized Beds
- Recent Advances in Fluid/Particle Separation
- Transport Phenomena in Fluid-Particle Systems
- Dynamics of Particulate Systems
- Micromechanics of Fluid Particle Systems
- Fluid Particle Systems in the Pharmaceutical Industry
- On-line Particle Sizing Methods
- Solids Flow Handling and Processing

The first article is an invited plenary paper by Dr. Ted Knowlton entitled "Unique Problems and Solutions in Gas-Solids Flow". This is a thought provoking review of some less than obvious aspects of standpipe flow.

I wish to acknowledge the chairs and co-chairs of the sessions for their careful review and selection of papers that resulted in the Chicago fluidization sessions being such a success. I would also like to acknowledge my co-editors, whose guidance has ensured that some of the best fluidization papers from the 1996 AIChE Annual Meeting were included in this volume. Finally, I would like to express my thanks to my administrative assistant, Gail Marais, without whose diligence this volume would not have been possible.

Desmond King, *Volume Editor*
General Manager
Alberta Envirofuels
P.O. Bag 2424, 9511 - 17th Street
Edmonton, Alberta T5J 4R3
Canada

CONTENTS

Foreword iii

PLENARY PAPER

Unique Problems and Solutions in Gas-solids Flow
Ted M. Knowlton 1

RESEARCH AND DEVELOPMENT PAPERS

Laser Doppler Anemometry (LDA) Diagnostics of Two Phase Flow Parameters in a
Liquid-Gas Two-Phase Flow System
S. Shao, G. Papavasiliou and H. Arastoopour 7

Liquid-Solid Fluidization Using Kinetic Theory
Dimitri Gidaspow and Lu Huilin 12

Discontinuity in Particle Granular Temperature Across the Geldart B/A Boundary
George D. Cody and David J. Goldfarb 18

Flow Behavior in the Riser of a High-Density Circulating Fluidized Bed
A.S. Issangya, D. Rai, J.R. Grace, K.S. Lim and J. Zhu 25

Solids Loading vs. Solids Flux as an Independent Parameter in CFB Similitude Studies
Ethan Brue and Robert C. Brown 31

A Detailed Experimental Description of the Flow in the Entrance Section of A FCCU
Apostolos E. Kostazos and Herbert Weinstein 36

Solids Residence Time Distribution in Interconnected Fluidized Beds
C.S. Stellema, Z.I. Kolar, J.J.M. de Goeij, J.C. Schouten and C.M. van den Bleek 40

Particle Clustering in Down Flow Reactors: Application of a Novel Fiber Optic Sensor
S. Krol, A. Pekediz and H. de Lasa 46

Gas-Liquid Mass Transfer in Three-Phase Inverse Fluidized Bed
Ihab H. Farag, Volen R. Nikolov and Iordan Nikov 51

Capacitance Probe Measurements of Solids Volume Concentrations and Velocities Inside an
Industrial Circulating Fluidized Bed Combustor
Bernd Hage and Joachim Werther 55

Hydrodynamic Aspects of Downflow Gas-Solids Reactors
N.K. Kimm, M.A. Forcinito and F. Berruti 61

A Modern Development of Friction Factors in Porous Media by Analogy to Turbulence
E.A. Stephan, M.S. Willis, R. Vengimalla and V.K.N. Yerra 67

A Volume-Average Scale Model for Fines Migration in Sandstone
E.A. Stephan and G.G. Chase 72

Simultaneous Sulfur and Metal Capture by Lime During Fluidized Bed Combustion
T.C. Ho, C.C. Shie, K.S. Wang and J.R. Hopper 77

Wall-to-Bed Heat Transfer in a Turbulent Fluidized Bed
Lii-ping Leu, Y.K. Hsia and C.C. Chen 83

Experimental Study of a Zigzag Air Classifier
X. Loison, A. Sahnoun, P. Guigon and J.F. Large 87

The Elastic and Cohesive Properties of Particulate Beds
A.J. Matchett, M. Aufauvre, J.M. Coulthard and S. Alsop 92

Axial Solids Distribution and Bottom Bed Dynamics for Circulating Fluidized
Bed Combustor Application
*Robert C. Zijerveld, Alain Koniuta, Filip Johnsson, Antonio Marzocchella, Jaap C. Schouten and
Cor M. van den Bleek* 97

Micromechanics of Particle-Particle and Fluid-Particle Interactions in Flows of Particulate Systems
R. Ocone 103

Magnetic Resonance Imaging of Granular Convection
James B. Knight, Heinrich M. Jaeger and Sidney R. Nagel 109

Modelling of Particle Interaction laws in Slow Shearing Granular Flows
D.M. Heyes, U. Tüzün and J. Baxter 113

Electrostatic Effects in Cold-Model Circulating Fluidized Beds
Charles J. Coronella and Jianxun Deng 119

A New Computer Controlled Wurster-type particle Coating Apparatus
H. Littman, M.H. Morgan III and C.B. Morgan 125

Characteristics of Particle Circulation in a Spouted Bed with a Draft Tube
Hongshen Ji, Atsushi Tsutsumi and Kunio Yoshida 131

Use of the Attenuation of Acoustic Pulsed Waves for Concentration Measurement in
Gas-Solid Pipe Flow
Stephen Tallon and Clive E. Davis 136

Characterization of Flow Properties of Very Fine Powders at Ambient and Elevated Temperature
[A Novel Experimental and Theoretical Approach]
H.O. Kono, L. Richman, J. Su and D. Smith 141

Local Bubble Properties in 40 and 80 cm Diameter Fluidized Beds
J.S. Groen, J.J. L. Berben, R.F. Mudde and H.E.A. Van Den Akker 147

A Numerical Model for Wave-Like Gas-Solids Flow in Pipes
David J. Mason and Shi Liu 152

Development of a Solids Flowmeter for an Industrial Scale Operation
John Kost, Jack Saluja, Jonathan Thorn, William Link and George Klinzing 156

Index 159

Plenary Paper
Fluor Daniel Lectureship Award

Unique Problems and Solutions in Gas-Solids Flow

Ted M. Knowlton
PSRI, 3424 S. State Street, Chicago, IL 60616

Two unique solutions to specific problems in gas-solid flow are described in this paper. In one of the solutions, the correct action is opposite to popular belief. In the other, the solution is unique or not expected. These unique situations are described below:

1. In most standpipes aeration is added to maintain the solids in fluidized flow. However, there are situations in which aeration is added to prevent fluidization from occurring.

2. In several instances, an L-valve has operated satisfactorily in a cold flow model, but was uncontrollable in the actual system. Why does this occur, and is it the fault of the L-valve?

In the past 15 years, the understanding of the transport of solids has increased significantly. Because of this understanding, there have been several rules-of-thumb and other generalizations that have become ingrained. However, as with almost any discipline there are "exceptions" to popular understanding. In this paper several situations are presented which appear to be contrary to what is normal, but in reality are quite logical when they are analyzed. Before analyzing the various problems and situations, it will be necessary to provide background on some of the principles of solids flow in both dilute and dense-phase.

PROBLEM 1 In many standpipes, aeration is added to prevent defluidization, or to control solids flow rates in non-mechanical valve systems. However, in one case of packed-bed standpipe flow, aeration was added to **prevent** fluidization from occurring. To understand why, it is necessary to understand standpipes and what happens when aeration is added to them..

The purpose of a standpipe is to transfer solids from a low pressure to a higher pressure. Therefore, pressure increases along the length of the standpipe. Fluidized-bed standpipes result in pressure increases along their length because of an increase in the "head" of the fluidized solids. In fluidized-bed standpipes operating at significant pressure ratios between the inlet and the exit, aeration is added to prevent the solids from defluidizing. Defluidization occurs because of the decrease in the gas volume of the interstitial gas as it transitions from low to high pressure. The decrease in gas volume causes the solids to become closer together and defluidize. Therefore, aeration is added to "replace" the gas volume decrease and to keep the solids fluidized. Fluidized standpipes are generally preferred to packed bed standpipes because they have less frictional resistance, and solids flux rates through the standpipe are higher.

In packed-bed standpipes, pressure buildup does not occur because of an increase in "head". Pressure buildup occurs because of the difference in relative gas velocity between the solids and the gas. The relative gas-solids velocity, v_r, is defined as:

$$v_r = |v_s - v_g|$$

where v_s is the velocity of the solids, and $v_g = U/\varepsilon$ is the interstitial gas velocity.

It is easier to visualize what is occurring in a solids transfer system by mentally traveling with the solids. Therefore, the positive reference direction for determining v_r is the direction that the solids are flowing. For standpipes, this direction is downward.

In Figure 1, solids are being transferred downward in a standpipe from pressure P_1 to a higher pressure P_2. Solids velocities in Figure 1 are denoted by the length of the thick-lined arrows, gas velocities by the length of the dashed arrows, and v_r by the length of the thin-lined arrows. In Case 1, solids are flowing downward, and gas is flowing upward relative to the standpipe wall. Therefore, v_r is directed upward, and is equal to the sum of v_s and the v_g, i. e.:

$$v_r = v_s - (-v_g) = v_s + v_g$$

For Case 2, solids are flowing down the standpipe. Gas is also flowing down the standpipe relative to the standpipe wall, but at a velocity less than that of the solids. For this case, v_r is also directed upward, and is equal to the difference between the solids velocity and the gas velocity, i. e.:

$$v_r = v_s - v_g$$

In both cases, if one were riding with the solids, the gas would appear to be moving upward.

Packed-bed standpipes generate a pressure increase along their length because of the difference in v_r between the gas and solids. In Figure 2A, solids are being transferred through an underflow packed-bed standpipe from the upper fluidized bed to the freeboard of the lower fluidized bed against the differential pressure P_2-P_1. The differential pressure P_2-P_1 consists of the pressure drop across the gas distributor of the upper fluidized bed ΔP_d and the pressure drop across the solids flow control valve ΔP_v. Therefore, the standpipe pressure drop ΔP_{sp} must equal the sum of ΔP_d and ΔP_v, i. e.:

$$\Delta P_{sp} = (\Delta P/L) H_{sp} = \Delta P_d + \Delta P_v = P_2-P_1 + \Delta P_v$$

Thus, for this packed-bed underflow standpipe case, the standpipe must generate a pressure drop greater than P_2-P_1. This is shown as Case I in the pressure diagram of Figure 2B.

If the gas flow rate through the column is increased, ΔP_d will increase. If ΔP_v remains constant, then ΔP_{sp} must also increase to balance the pressure-drop loop. This is shown as Case II in the pressure diagram of Figure 2B. Unlike an overflow standpipe, the solids level in the standpipe cannot rise to increase the pressure drop in the standpipe. However, the $\Delta P/L$ in the standpipe must increase in order to balance the pressure drop around the loop. This occurs in a packed-bed standpipe because of an increase in v_r in the standpipe. This can be visualized with the aid of Figure 2C.

For Case I, the ΔP in the bed was satisfied by having the standpipe operate at point I on the $\Delta P/L$ versus v_r curve (Figure 2C). When the ΔP across the distributor increased, the v_r in the standpipe adjusted to generate a higher $\Delta P/L$, $(\Delta P/L)_{II}$, to balance the higher pressure drop. If the ΔP across the distributor is increased such that the product of $(\Delta P/L)_{mf}$ and the standpipe length, H_{sp}, is less than the sum of ΔP_v and ΔP_d, then the underflow standpipe will not seal.

In one standpipe application, a two-component particulate solid was to be transferred in a standpipe over a pressure range of several bar. The solids were a heavy solid with an average particle size of about 100 microns, and a less-dense solid with an average particle size of nearly 1000 microns. Operating in the fluidized bed mode, the smaller solids segregated from the larger solids. This lead to severe slugging in the standpipe as shown in Figure 3. The solution to the problem was to add aeration to the standpipe to prevent fluidization from occurring. Without aeration, the upper part of the standpipe was in packed bed flow, but the lower part of the standpipe became fluidized and slugged. Diagrams of v_r and the pressure-drop-per-unit- length ($\Delta P/L$) in the standpipe are also shown in Figure 3. v_r increased from the top of the standpipe to the bottom because increasing pressure caused the gas volume to decrease. At some point in the standpipe, the solids became fluidized, segregation occurred, and the solids began to slug. To prevent slugging, aeration was added to the standpipe to decrease v_r and solve the slugging problem. The change in v_r and $\Delta P/L$ along the standpipe for aeration addition at three points in the standpipe is shown schematically in Figure 4.

PROBLEM 2 *To determine if an L-valve would control solids in one process, the solids (180 microns in size) were tested in a cold-flow L-valve1. The L-valve was to control the flow rate of solids from a fluidized bed into a pneumatic conveying line. Tests showed that the L-valve controlled the solids well, and it was designed into the process. However, during shakedown in the commercial equipment, the solids could not be controlled by the L-valve. Why?*

To determine why, it is necessary to understand what causes solids to flow in an L-valve. The L-valve is one of a class of valves called non-mechanical valves which use only aeration gas in conjunction with their geometrical shape to control the flow rate of solids. Therefore, the solids flow rate through an L-valve is controlled by the amount of aeration gas added to it. When aeration gas is added to an L-valve, gas flows downward through the particles and around the constricting bend. This relative gas-solids flow produces a frictional drag force on the particles. When this drag force exceeds the force required to overcome the resistance to solids flow around the bend, solids flow through the L-valve. The gas flow that causes the solids to flow around the L-valve is not the amount of aeration gas added to the valve. If gas is traveling down the moving packed-bed standpipe (which occurs in most cases) with the solids, the amount of gas which flows around the L-valve bend, Q_T, is the sum of the standpipe gas flow, Q_{sp}, plus the aeration gas flow, Q_A, (Figure 5A). If gas is flowing up the standpipe (this occurs when the standpipe is operating with low solids flow rates and/or large solids), then the amount of gas flowing around the bed, Q_T, is the difference between the aeration gas flow and the gas flowing up the standpipe as shown in Figure 5B.

When an L-valve is used to control solids flow, it is always located at the bottom of an underflow standpipe operating in moving packed-bed flow. The standpipe is generally fed by a non-fluidized hopper or a fluidized bed. Knowlton and Hirsan (1) have shown that the operation of a nonmechanical valve is dependent upon the pressure balance and the geometry of the system.

The cold-flow L-valve control system used to test the solids in the problem is shown in Figure 6A. The high-pressure point in this system was at the L-valve aeration point. The low-pressure common point was in the freeboard above the fluidized bed. The pressure-drop balance around the cold model was:

$$\Delta P_{L\text{-valve}} + \Delta P_{riser} = \Delta P_{sp} + \Delta P_{fluid\ bed}$$

The bed height remained essentially constant, so that the equation above can be re-written as:

$$\Delta P_{L\text{-valve}} + \Delta P_{riser} - K = \Delta P_{sp}$$

where K is the constant ΔP across the fluidized bed. For the cold model, the fluidized bed was short and the pressure drops across the L-valve and the riser were greater than that across the fluidized bed, i.e.,

$$\Delta P_{L\text{-valve}} + \Delta P_{riser} \gg K$$

Therefore, for the cold model, ΔP_{sp} was positive, and the cold-model standpipe built pressure in order for the pressure drops around the loop to balance. The solids could be started, stopped, and controlled using the L-valve. However, the solids in the commercial L-valve could not be controlled. In fact, they could not be stopped from flowing. What was the problem?

A schematic drawing of the commercial unit and the pressure-drop loop is also shown in Figure 6B. The pressure balance around this system is exactly the same as that of the cold model, i. e.,

$$\Delta P_{L\text{-valve}} + \Delta P_{riser} - K = \Delta P_{sp}$$

However, the fluidized bed in the commercial system was so tall that the ΔP across it was greater than the ΔP across the L-valve and the riser. Therefore, the ΔP across the standpipe was negative. This meant that the standpipe had to dissipate pressure instead of build it, so that the pressure at the top of the standpipe was greater than the pressure at the bottom of the standpipe. For this to occur, v_g in the packed-bed standpipe had to be greater than v_s. In the commercial unit v_r was great enough to drag solids around the bend of the L-valve, and the L-valve would not shut off. It was not the "fault" of the L-valve, but a lack of understanding of how the L-valve worked and inadequate design which resulted in the inability of the L-valve to stop the solids - even with the L-valve aeration off. To make the L-valve work in this situation, the entrance pipe for the L-valve had to be moved up into the fluidized bed. Then, the pressure at the top of the standpipe was not great enough to cause negative pressure drops across the standpipe.

The commercial system was operated at a temperature of about 1000°F. L-valves operated at elevated temperature require less aeration gas (a smaller v_r) to cause solids to flow. The reason for this is explained below. The ΔP produced by gas flowing through the solids in a packed bed is approximated by the Ergun Equation, which is of the form:

$$\frac{\Delta P}{L} = K_1 \mu v_r + K_2 \rho_g v_r^2$$

The second term in this equation is the pressure drop due to kinetic forces, while Term 1 is the pressure drop due to viscous forces. For most packed beds, Term 2 is small and can be neglected. Therefore:

$$\frac{\Delta P}{L} = K_1 \mu v_r$$

At elevated temperature, the viscosity of the gas increases. This means that the same $\Delta P/L$ can be produced by a lower v_r. For the same solids flow rate, this means that a smaller gas flow is required to produce the same frictional pressure drop, in the L-valve. Therefore, at temperature, less gas flow was required to cause the solids to flow through the L-valve than at ambient conditions. Because gas viscosity increases by about a factor of 2 as the temperature is raised from ambient to about 1500°F, at temperature, the gas flow through the L-valve required to cause solids flow is about half that at ambient conditions.

LITERATURE CITED

1. **Knowlton, T. M. and Hirsan,** I. "L-valves characterized for solids flow", *Hydrocarbon Processing,* **57**, pp. 149-159 (1978).

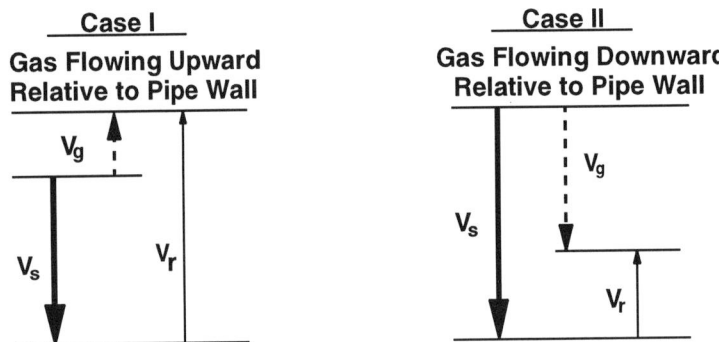

Figure 1. Concept of Relative Gas-Solids Velocity

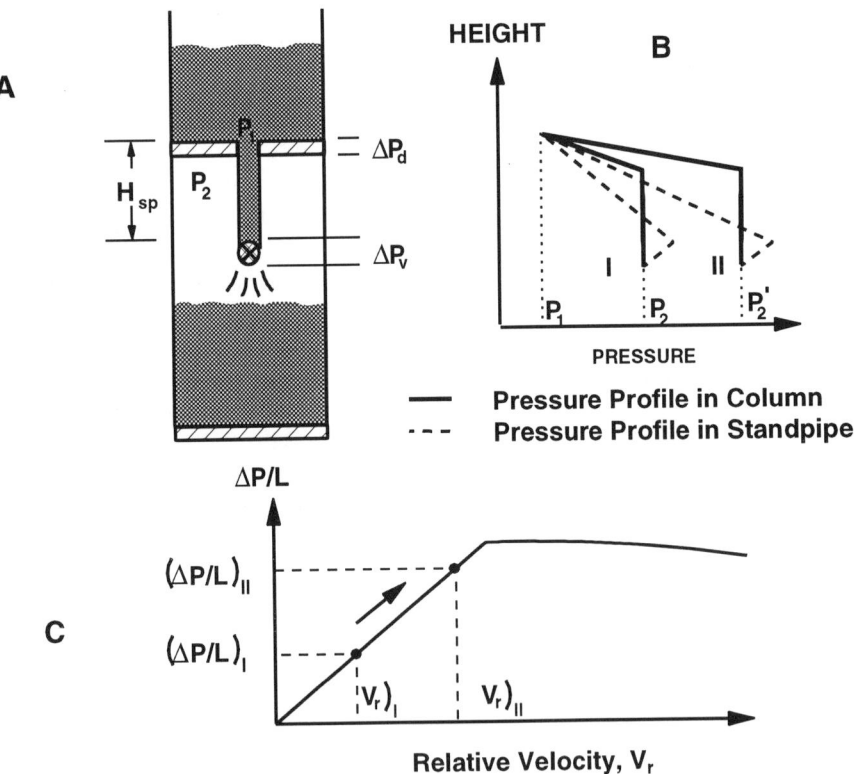

Figure 2. Underflow Standpipe Operation

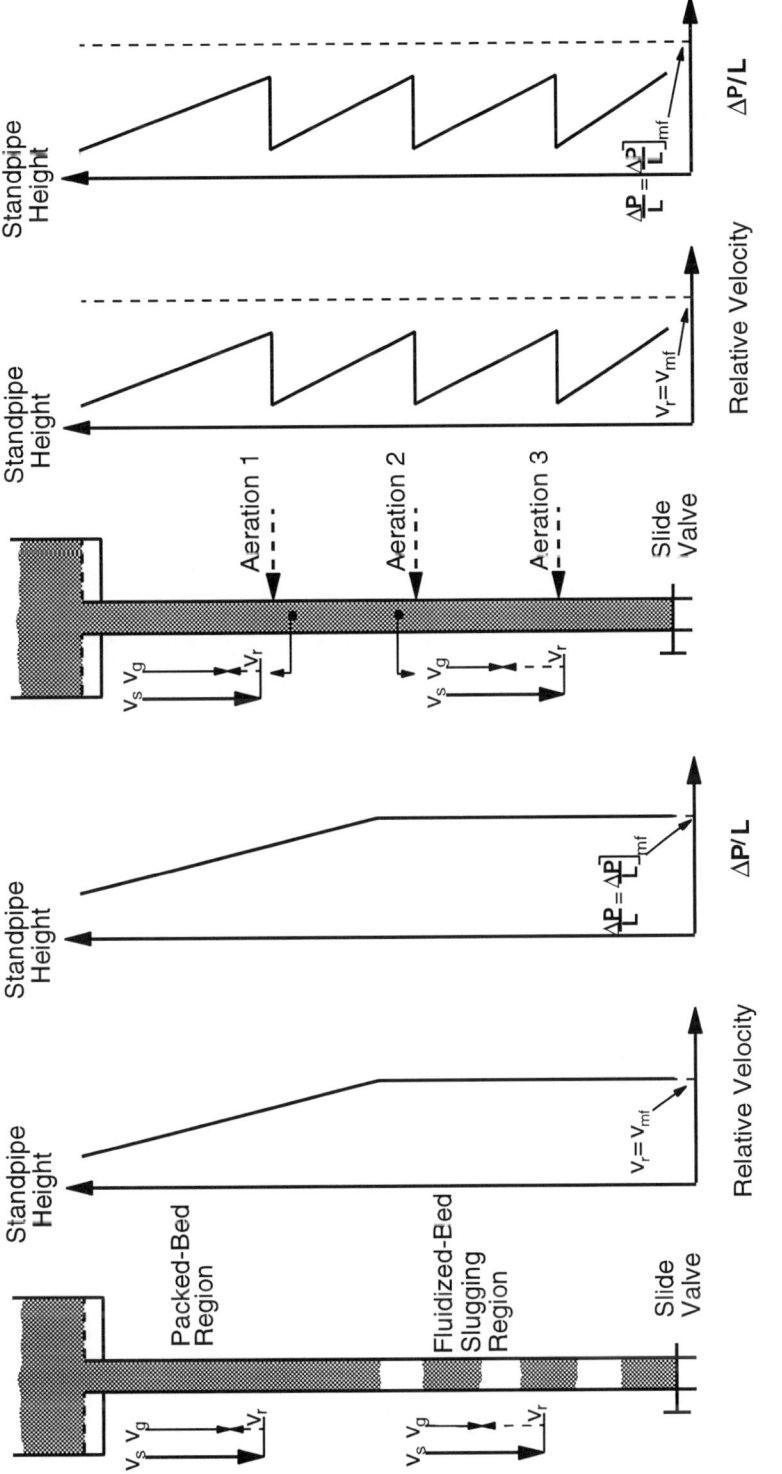

Figure 4. Standpipe Relative Velocity and Pressure Drop Profiles (Aeration)

Figure 3. Standpipe Relative Velocity and Pressure Drop Profiles (No Aeration)

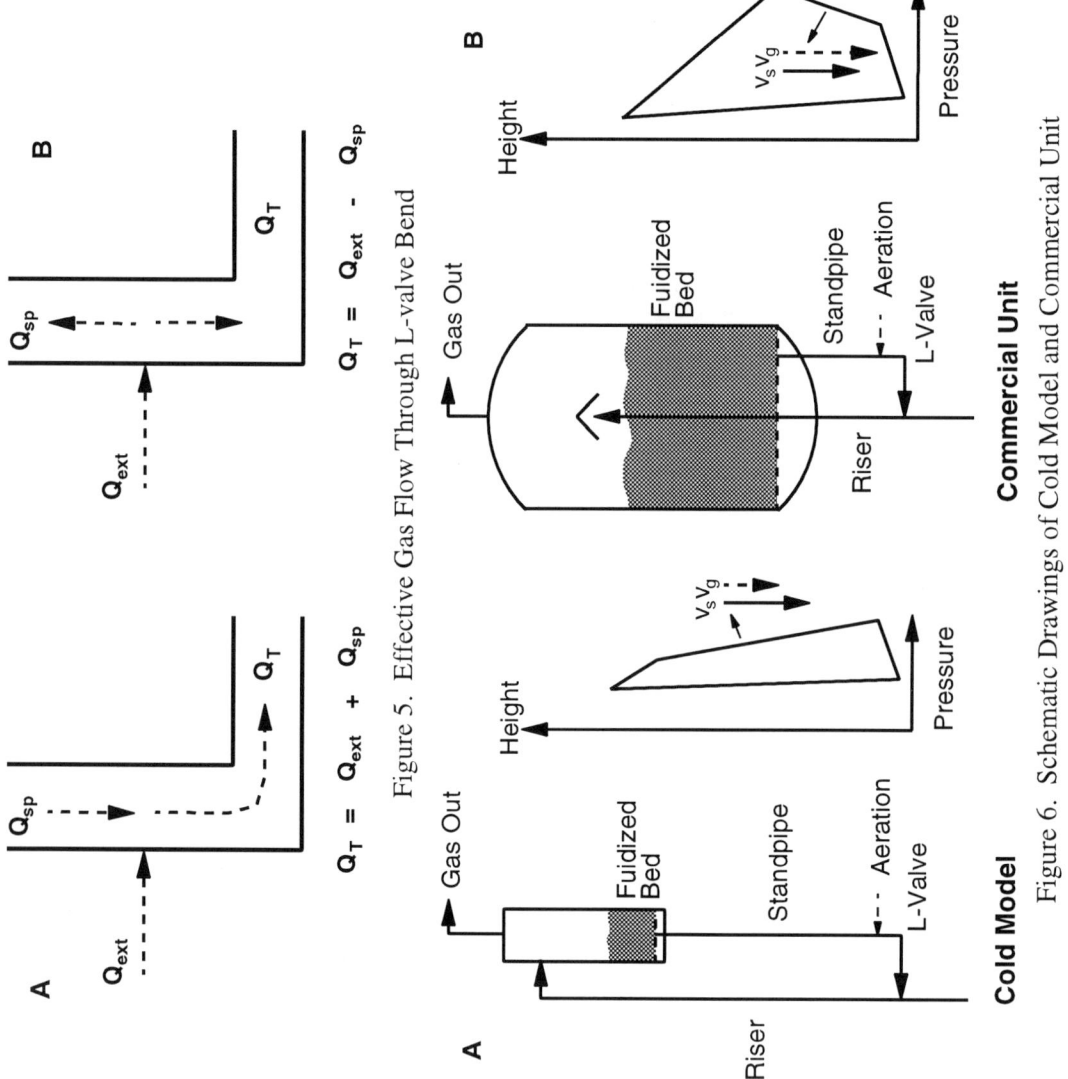

Figure 5. Effective Gas Flow Through L-valve Bend

Figure 6. Schematic Drawings of Cold Model and Commercial Unit

Laser Doppler Anemometry (LDA) Diagnostics of Two Phase Flow Parameters in a Liquid-Gas Two-Phase Flow System

S. Shao, G. Papavasiliou and H. Arastoopour
Department of Chemical & Environmental Engineering, Illinois Institute of Technology, Chicago, IL 60616

A Laser Doppler Anemometry (LDA) technique has been developed for the measurements of two phase flow parameters under the coexistence of two phases in a liquid-gas two-phase flow system. Based on the principle of geometric optics, both the forward and the backward scattering modes are adopted for the receiving optics. The threshold technique is used for the backward scattering mode to suppress the signals generated by the liquid phase and to make sure the signal processor can be triggered only by the signals generated by bubbles or solid particles. When bubbles or solid particles are large, the liquid-gas or the liquid-solid interface will act mainly as a reflecting mirror. This will make the forward receiving optics sitting on the optical axis receive the light scattered by the tracers only. This makes it possible to measure the liquid phase velocity under the coexistence of two phases.

This technique has been applied to a liquid-gas two-phase upward flow system to measure dynamically two phase velocities, bubble size distributions and bubble number of density. The turbulence intensities of both phases are calculated and compared with that of single phase liquid so that the effect on the liquid phase due to the presence of bubbles can be studied. The experimental results show that this technique is successful for the measurements of two phase flow parameters under the coexistence of two phases in a liquid-gas flow system.

Two-phase flow systems, such as liquid-gas, have significant applications in many industrial processes. Experimental studies on the flow behavior of bubbles in a liquid-bubble two-phase flow system are of great interest due to their applications in the chemical, waste treatment and bioprocessing industries. Thus, experimental results obtained under two-phase flow conditions play a critical role in the optimum design of such systems.

For gas-solid and air-droplet two-phase flow systems, it has been found that the presence of solid particles and droplets has significant impact on the flow patterns, turbulence, and heat and mass transfer. For example, Hishida et al [1] studied turbulent flow characteristics of dispersed two-phase flows and found the suppression of the Reynolds shear stress due to the presence of solid particles in a dispersed gas-solid two-phase flow. For liquid-bubble two-phase flow systems, on the other hand, there are no significant experimental data available in the literature regarding the flow parameters of the two phases which were measured under the two-phase flow conditions.

In the present study, a special orientation of the receiving optics was adopted so that both the bubble and the water flow parameters were measured under two-phase flow conditions. Comparisons of the flow parameters between the single-phase water and the water phase were made in order to observe the impact of the bubbles on the flow behavior of the liquid phase.

MEASURING TECHNIQUE

The technique used in this study is based on the principle of geometric optics and incorporates both the forward and the backward scattering modes. The threshold technique was used for the backward scattering mode to suppress the signals generated by the liquid phase and to make sure the signal processor could be triggered only by the signals generated by the bubbles. When bubbles are large, the liquid-gas interface acts as a reflecting mirror which allows only the light scattered by the tracer to be received by forward receiving optics. Thus, the liquid and the bubble flow parameters can be obtained respectively from the forward and backward receiving optics. This technique requires two sets of receiving optics and signal processors. For a steady state flowing system, one set of receiving optics and signal processor may be used. In addition, the bubble diameter must be larger than the size of the measuring volume.

EXPERIMENTAL SET-UP AND PROCEDURE

The flow parameters were measured for the bubbles and the water in a water-bubble two-phase upward flow system which is shown schematically by Figure 1. The system consisted of an in-line pump, a glass tube which was aligned vertically, a storage tank, and two pieces of connecting tubing A and B, which transferred water from the glass tube to the storage tank and connected the in-line pump to the glass tube, respectively. Bubbles were generated by air entrainment due to the water jet entering the free surface of the water in the tank. The water jet was formed at the exit of tubing A.

Before bubbles appeared in the system, the velocity profile along the radial position was measured for the single-phase water flow by measuring tracer particle velocity in the forward scattering mode. The backward scattering mode was used to measure bubble velocity and size distribution in the radial direction. The gain of the signal conditioner was reduced to suppress tracer signals. To be sure that the signals generated by even the smallest bubbles in the system were detected, the gain was set at an optimum value such that the signal processor was not triggered by the tracers. This ensured that only signals with sufficient amplitude, such as those generated by bubbles, were able to trigger the signal processor. After careful adjustment of the settings, the exit end of tubing A was raised above the free surface of the water in the tank. The distance between the exit end of tubing A and the free surface of the water in the tank was adjusted to control the amount of air that was entrained into the system and the size of the bubbles that were generated. In this study, two jet lengths were used: 0.11" and 0.22", corresponding to the calculated entrained air volume of 377 cm^3/s and 755 cm^3/s, respectively [2].

EXPERIMENTAL RESULTS AND DISCUSSIONS

Figure 2 compares the velocity profiles of the bubbles, the water and the single-phase water. Unlike the liquid-solid two-phase flow system in which the density of the particulate phase is higher than that of the carrying liquid and the inertia effect makes the particulate phase lag behind the liquid phase, the water-bubble flow system exhibits an opposite trend. The liquid phase lags behind the bubble phase as shown clearly by Figure 2. The buoyancy effect causes the bubbles to flow faster than the water.

The presence of the bubbles affected the water velocity profile considerably. For example, in the presence of bubbles, the maximum velocity of the water phase at the central part of the tube rose from around 0.30 m/s to 0.32 m/s, a 7% increase. In the regions away from the central part of the tube, the water phase velocity decreased as compared with the velocity of the single-phase water. The maximum reduction was 0.02 m/s (a 7% decrease), and occurred at the point 4 mm away from the wall. Therefore, it can be summarized that the presence of bubbles increased the water phase velocity at the central part of the tube and reduced it at the other regions of the tube.

The bubbles, particularly the larger ones, have tendency to move to the higher velocity region, i.e., the central part of the tube, due to the Magnus effect. Since bubbles moved faster than water, they transferred momentum to the surrounding water and caused an increase in the velocity of water phase at the center of the tube (see Figure 2). Under the steady state operation, due to the conservation of mass for the water phase, the velocity of the water decreased at the other regions of the tube. This interpretation is confirmed by Figure 3 which shows the velocity distribution of the single-phase water and water phase at different bubble concentrations. Higher bubble concentrations enhanced the increase in the water phase velocity at the center of the tube and also enhanced the decrease in the water phase velocity at the other regions of the tube. At a very low bubble concentration, the variation in the water phase velocity compared with the single phase water flow was negligible.

Figure 4 shows the bubble size distribution at two different bubble concentrations corresponding to two different water jet lengths. As shown, bubble size increased when the water jet was longer (higher bubble concentration), which is consistent with correlation in Perry et al Handbook [2]. Also shown by Figure 4, larger bubbles appeared in the center of the tube due to the Magnus effect.

Figure 5 shows the comparison between the turbulence intensity distribution of the bubble and water phases. Both phases showed higher turbulence intensity in the wall region than in the central region of the tube, which is believed to be caused by the lower velocity at the wall. On the other hand, the bubble exhibited a stronger turbulent behavior than the water phase. At most of the measured points, particularly those close to the wall of the tube, the bubbles showed a higher turbulence intensity. The maximum turbulence intensity of the bubbles was around 0.15 and the maximum turbulence intensity of the water phase was around 0.08. Both occurred at same location close to the wall.

It is revealing and interesting to compare the turbulence intensity of the water phase with that of the single-phase water to observe how the presence of the bubbles affect the turbulent behavior of the water phase. As shown by Figure 6, single-phase water showed higher turbulence at the wall region and lower turbulence at the central region. When bubbles appeared in the system, this pattern of turbulence changed: the turbulence intensity at the wall region decreased. As the concentration of the bubbles increased and larger bubbles were generated in the system, the turbulence intensity decreased further (from 0.14 for the single-phase water to 0.9 for the water phase, a 36% reduction). Such drastic reduction in turbulence intensity at the wall region could cause problems in designing some engineering processes involving heat transfer. A decrease in turbulence intensity may significantly reduce the heat flux [3].

CONCLUSION

Based on the experimental results, the following conclusions may be drawn: (1) The turbulence intensity of the bubbles was higher than that of the water phase at the wall region. The bubbles and the water phases had the same level of turbulence at the central region; (2) The velocity of the bubbles was higher than that of the water phase. This phenomenon was more pronounced at higher bubble concentrations; (3) When the system was operated at higher bubble concentration, the presence of bubbles caused an increase in the mean velocity of the water phase at the central region, and a decrease in the mean velocity of the water phase at the wall region of the tube and (4) The presence of bubbles significantly affected the turbulence pattern of the water phase. The turbulence intensity of the water phase was appreciably hindered at the wall region, and slightly changed at other regions away from the wall.

LITERATURE CITED

1. **Hishida, K., Ando, A., Hayakawa, A. and Maeda, M.** "Turbulent Flow Characteristics of Dispersed Two-Phase Flow in Plane Shear Layer," *Applications of Laser Anemometry to Fluid Mechanics,* Eds.: R.J. Adrian, T. Asanuma, F. Durst, J.H. Whitelaw, Spring-Verlag (1989).

2. **Perry, R.H., Green, D.W. and Maloney, J.O.** *Perry's Chemical Engineers' Handbook*, Sixth Edition, McGraw-Hill Book Company (1984).

3. **Eckert, E.R.G. and Drake Jr., R.M.** *Analysis of Heat and Mass Transfer*, McGraw-Hill Book Company (1972).

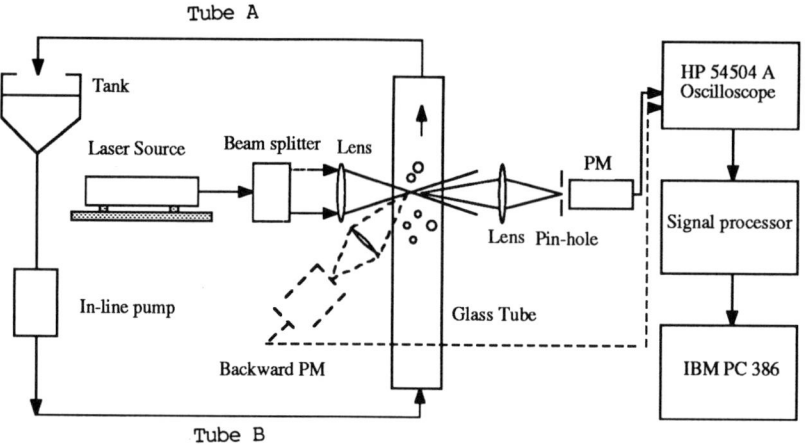

Figure 1 Schematic Diagram of a Water-Bubble Two-Phase Flow and Measuring System

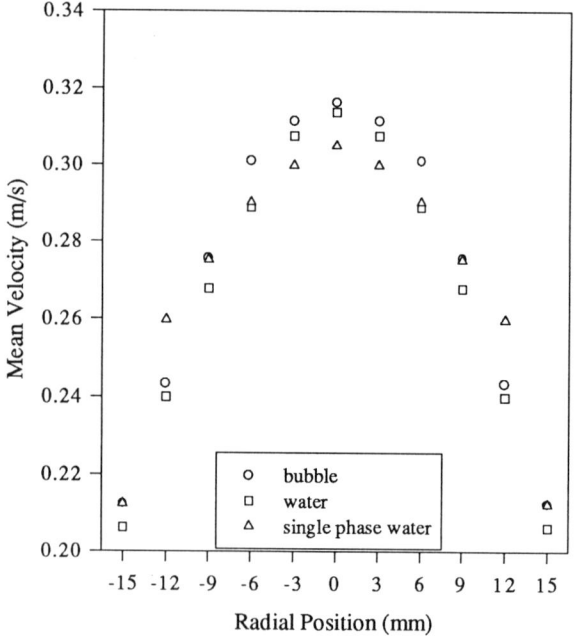

Figure 2 Mean velocity distributions of the single phase water, the bubbles and the water phase in a liquid-gas flow system.

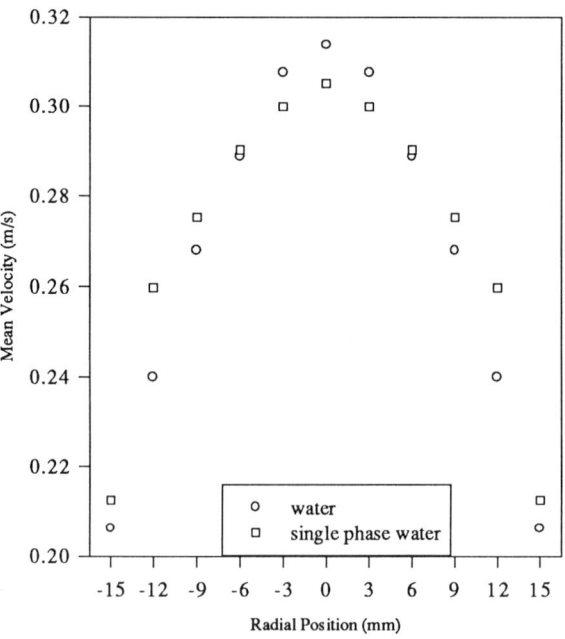

Figure 3 Mean velocity distributions of the single phase water, the bubbles and the water phase in a liquid-gas two-phase flow system.

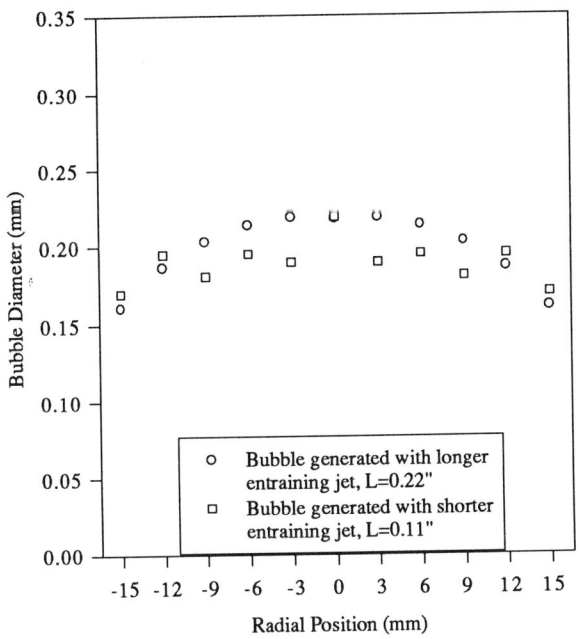

Figure 4 Bubble size distributions at different lengths of the entraining jet in a liquid-gas two-phase flow system

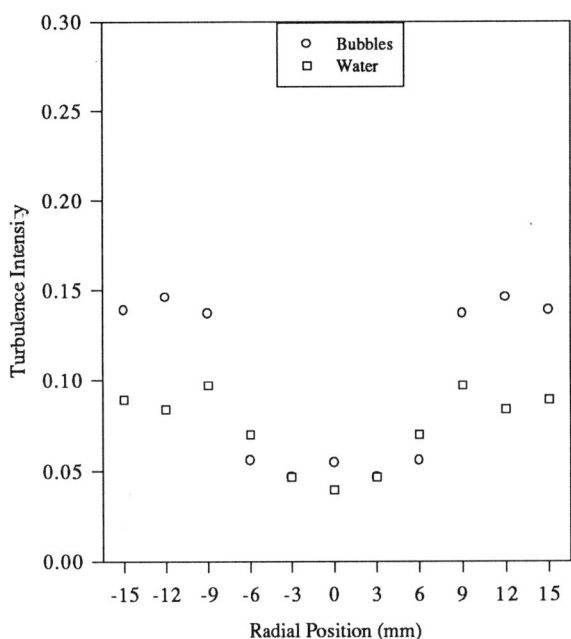

Figure 5 Turbulence intensity distributions of the bubbles and water in a liquid-gas two-phase flow system (Entraining jet length = 0.22").

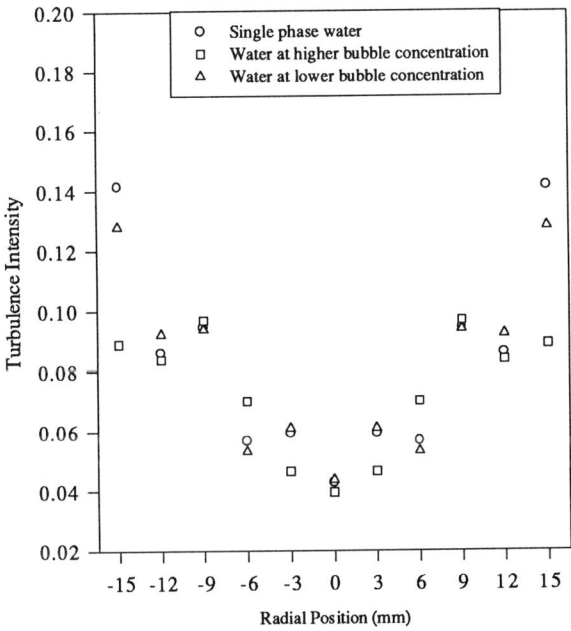

Figure 6 Turbulence intensity distributions of the single phase water and the water phase at different bubble concentrations in a liquid-gas two-phase flow system

Liquid-Solid Fluidization Using Kinetic Theory

Dimitri Gidaspow and Lu Huilin
Department of Chemical and Environmental Engineering
Illinois Institute of Technology, Chicago, IL 60616

The granular temperature of Air Products catalyst particles and 450 µm glass beads were measured in a two dimensional fluidized bed using our CCD camera technique used previously for measuring the viscosity of FCC particles. Similarly to the FCC particles in CFB the dominant frequency of the catalyst particles in water correlated with the sonic velocity of the Air products catalyst.

For 450 µm glass beads our camera technique allowed us to measure the radial distribution function of statistical mechanics. It showed us that the particles never contacted each other during collisions. The values of the radial distribution function determined from the first peak lie between those obtained from the Bagnold equation and the hard sphere model. These values are used in determining the particulate viscosities. For the glass beads these viscosities agreed with the measurements using a Brookfield viscometer.

To better understand the dynamics of slurry bubble column reactors, we studied the hydrodynamics of fluidization of 450 micron glass beads and 50 micron catalyst particles used to make methanol from syn-gas.

FLUIDIZATION EQUIPMENT

A two-dimensional bed was used as the fluidization unit. See Fig.1. Its total height was 1.85 m, with a 1.2 m high test section. The cross-section was 30.48 cm by 5.08 cm. A pump was connected to the bottom of the bed by a 2.54 cm I.D. stainless steel pipe. A water distribution system consists of a pipe distributor containing 42 holes (d=4 mm) and a perforated plate. A plastic grid was attached to the top of the bed to minimize the entrainement of particles by water. The particles were fed into the bed by means of a screw feeder placed on top of the bed to maintain constant solid weight in the bed. Copper oxide catalyst particles with an average diameter of 50 µm and a density of 3000 kg/m³ and glass beads with an average diameter of 450 µm and a density of 2600 kg/m³ were used as the solid phase. Still bed height of particles was 20 cm from the perforated distributor. Local solid volume fractions were measured by means of a conductivity probe and the particle velocities were determined using our high resolution micro-imaging/measuring system which was previously used to measure the particle velocity in the three phase fluidized bed (Gidaspow et al., 1995$_a$).

Particle Velocity Determination

The high resolution micro-image/measuring system is essentially a 2/3 inch color video camera (DXC-151A) which uses a Charge Couple Device (CCD), a solid stage sensor. The camera has several electronic shutter settings for the particle velocities measurements. The personal computer which has a Micro-Imaging Board and a Micro-imaging software stored the captured images for further analysis (Gidaspow, et al., 1995$_a$; Gidaspow & Huilin, 1996).

Figure 2 shows typical streaks made by the particles recorded by the High Resolution Micro-Imaging/Measuring System. These streak lines represent the distance traveled by the particles in a given time interval specified on the camera. From these streaks, particle speed and velocities can be determined.

Particle Concentration Measurements

Local solid volume fraction measurements were performed with a conductivity technique (Turner, 1973; Begovich & Watson, 1978). A pair of brass electrodes, 1.6 cm height and 1.1 and 1.3 cm width, were placed at the center of the wall opposite each other at a height of 10 cm from the plate distributor.

The small electrode thickness (1.0 mm) did not modify local hydrodynamics. The probe combining a copper-constant thermocouple connected into a 1481-00 conductivity meter (Cole-Parmer Instrument Company). Its signal connects to a T31-B terminal (Strawberry Tree), and is sampled by an IBM PC computer with a data acquisition card, ACAO mounted in the computer, by a data acquisition software package called ACQ. Sampling frequency was at least 200 Hz with a sampling time of 30 seconds. This yielded the most reproducible and stable results. A calibration curve between the solid volume fraction and the voltage from the conductivity meter was obtained. This calibration was found to give a reasonable resolution when the solid volume fraction was greater than about 0.3 in this study.

PART I: 50 μm CATALYST PARTICLES

Solid Volume Fraction Fluctuations

Typical solid volume fraction fluctuations with time are shown in Figure 3 for a liquid velocity of 0.77 cm/s. The time-averaged solid volume fraction corresponding to this liquid velocity was 0.5397. Figure 4 shows the power spectrum density of solid volume fraction fluctuations at this liquid velocity. A dominant peak is present. Figure 5 shows the dominant frequency as a function of solid volume fraction. The dominant frequency increases with an increasing liquid velocity, as in the study of Anderson & Jackson (1969).

Particle Velocity and Granular Temperature

Measured particle speed and the calculated granular temperatures are summarized in the Table 1. The granular temperature is one third of the sum of the deviations of particle oscillations in the three directions. The standard deviations, σ_x and σ_y were calculated based on instantaneous velocity of particles from the mean values:

$$\theta = \frac{1}{3}\left(\sigma_x^2 + \sigma_y^2 + \sigma_z^2\right) = \frac{1}{3}\sigma_x^2 + \frac{2}{3}\sigma_y^2 \qquad (1)$$

and

$$\sigma = \sqrt{\frac{1}{N}\sum_{i=1}^{N}(u_i - u_m)^2} \qquad (2)$$

where u_i and u_m are the instantaneous and the mean particle velocity, respectively.

Particulate Sonic Velocity

The pseudo-sonic velocity of particles is (Gidaspow, 1994):

$$C_s = \sqrt{\frac{1}{\rho_s}\frac{\partial p_s}{\partial \varepsilon_s}} \qquad (3)$$

where

$$p_s = \rho_s \varepsilon_s \theta\left[1 + 2\varepsilon_s g_0(1+e)\right] \qquad (4)$$

Figure 6 shows that there exists a link between the major frequency of particle oscillations and the sonic velocity of particles determined from the granular temperature, similar to that reported for FCC particles in air at FLUIDIZATION VIII and predicted from theory of wave propagation (Gidaspow et al. 1995$_b$). The wave length determined in figure 6 is 4.32 cm. It is roughly the distance between the plates. Hence the third dimension is important in this system.

Reynolds Stresses Computation

From the particle velocity measurements, we can compute the Reynolds stress as follows:

$$<u'v'> = \frac{1}{N}\sqrt{\sum_{i=1}^{N}(u_i v_i - \overline{uv})^2} \qquad (5)$$

where $\overline{uv}$ is the mean value of particle velocities. Figure 7 shows the computed Reynolds stress as a function of solid volume fraction. We can see that the Reynolds stress increases with decreasing solid volume fraction. The trend of the Reynolds stress is the same as the measured granular temperature behavior listed in Table 1.

PART II RADIAL DISTRIBUTION FUNCTION FOR 450 μm GLASS PARTICLES

For determination of collisional viscosity we used the equation (Gidaspow, 1994):

$$\mu_s = \frac{5\rho_s d_p \sqrt{\pi\theta}}{48(1+e)g_0}\left[1 + \frac{4}{5}(1+e)g_0\varepsilon_s\right]^2 + \frac{4}{5}\varepsilon_s^2 \rho_s d_p(1+e)\left(\frac{\theta}{\pi}\right)^{0.5} \qquad (6)$$

where g_0 is the radial distribution function at contact, and e is the restitution coefficient. A simple form for the radial distribution function g_0 was that used by Bagnold (1954):

$$g_o = \left[1 - \left(\varepsilon_s / \varepsilon_{s,\max}\right)^{1/3}\right]^{-1} \quad (7)$$

In this study we determined the accuracy of the Bagnold equation. In statistical mechanics of liquids the radial distribution function measures the local density. We found that our CCD camera system can easily measure this quantity, since the coordinates of the particles are stored in the computer system. Table 2 shows how we measured this density. Previously direct measurements of the radial distribution function of colloidal particles have been carried out using a camera method (Ise et al., 1983; Yoshida et al., 1990). Figure 8 shows a typical particle distribution image captured by our CCD camera. The coordinates of each particle can be determined by the software IPPLUS. Hence the variation of the radial distribution function with distance can be found (Egelstaff, 1967; Hunter, 1989; Balucani & Zoppi, 1994). Figure 9 shows a typical calculated radial distribution function profile. From this plot, the radial distribution function at contact g_o was obtained. It is the value of the first peak. Unlike in the gas-solid system, this peak does not occur at particle contact r = d. It is almost 50 % larger. This is due to the fact that the particles start to fly apart before contact. There is a film between the particles. It gives rise to the lubrication force (Hunter, 1989). Figure 10 shows the variation of the radial distribution function g_o with solid volume fraction. We see that the Bagnold equation we used previously agrees with the data.

Table 3 summarized the measured particle speed and the calculated granular temperatures of 450 µm glass beads in the liquid-solid fluidized bed. The glass beads concentration was measured by means of the conductivity technique.

Figure 11 shows that the viscosity computed from kinetic theory agrees with the Brookfield viscometer measurements for 450 micron particles. Similar agreement was obtained by Bahary (1994) for 800 micron particles in a G-L-S fluidized bed. However, the Brookfield viscometer measurements for the 50 micron catalyst particles give a viscosity 10 times higher than those in figure 11. We believe this is due to the much higher values of g_o for the catalyst mixture.

ACKNOWLEDGMENT:
This study was supported by DOE grant No: DE-FG-94PC94208 and NSF grant CTS-9305850.

NOMENCLATURE:
C_s sonic velocity
d_p particle diameter
e   restitution coefficient
g_o radial distribution function at contact
g(r)   radial distribution function
N   total number of sampling
P_s particulate pressure
u_i instantaneous particle velocity
u_l liquid velocity
u_p averaged particle velocity
<u'v'>   Reynolds stress
v_i instantaneous particle velocity
uv   mean value of particle velocities
r   distance
θ   granular temperature
σ   standard deviation
ρ_s particle density
ε_s solid volume fraction
$\varepsilon_{s,\max}$ solid volume fraction at packing
μ_s particulate viscosity

LITERATURE CITED:
1. Anderson T. B. and R. Jackson, "Fluid Mechanical Description of Fluidized Beds, Comparison of Theory and Experiment", *Indust. Engng Chem. Fundls*, **8**, 137-144 (1969).
2. Bagnold R. A., "Experiments on a Gravity-free Dispersion of Large Solid Spheres in a Newtonian Fluid under Shear", *Proc. Roe. Soc.*, **A255**, 49-63 (1954).
3. Bahary, M., "Experimental and Computational Studies of Hydrodynamics in Two and Three Phase Fluidized Beds", Ph. D. Thesis, IIT (1994).
4. Balucani U. and Zoppi M., "Dynamics of The Liquid State", Oxford Science Publications (1994).
5. Begovich J. M. and J. S. Watson "An Electroconduc-tivity Technique for Measurements of Axial Variation of Holdups in Three-phase Fluidized Beds", *AIChE. J.*, **24**, 351-354 (1978)
6. Carnahan N. F. & K. E. Starling, "Equation of State for Nonattracting Rigid Spheres", *J. Chem. Phys.*, **51**, 635-636 (1969).
7. Didwania A. K. and G. M. Homsy, "Flow Regimes and Flow Transitions in Liquid Fluidized Beds", *Int. J. Multiphase Flow*, **7**, 563-580 (1981).
8. Egelstaff P. A., "An Introduction to The Liquid State ", Academic Press (1967).

9. Gidaspow D., M. Bahary and W. Yuanxiang, "Hydrodynamic Models for Sluury Bubble Column Reactor", *Fluidization & Fluid-particle Systems*, Preprints of 1995 AIChE Annual Meeting, Florida, 164-168 (1995$_a$).

10. Gidaspow D., L. Huilin and A. Therdthianwong, "Measurement and Computation of Turbulence in a Circulating Fluidized Bed", *Fluidization VIII*, Edited by C. Laguerie & J. F. Large, 81-88 (1995$_b$).

11. Gidaspow D. and L. Huilin, "Collisional Viscosity of FCC Particles in a CFB", *AIChE. J.*, **42**, 2503-2510 (1996).

12. Gidaspow D, "Multiphase Flow and Fluidization: Continuum and Kinetic Theory Descriptions", Academic Press (1994).

13. Hunter R. J., "Foundation of Colloid Science II", Oxford Science Publications, Calarendon Press (1989).

14. Poletto M., R. Bai and D. D. Joseph, "Propagation of Voidage Waves in a Two-dimensional Liquid-Fluidized Bed", *Int. J. Multiphase Flow*, **21**, 223-239 (1995).

15. Ise N., T. Okubo, M. Sugimura, K. Ito & H. J. Noite, "Order Structure in Dilute Solutions of Highly Charged Polymer Lattices as Studied by Microscopy I. Interparticle Distance as a Function of Latex Concentration", *J. Chem. Phys.*, **78**, 536-540 (1983).

16. Turner J. C. R., "Electrical Conductivity of Liquid Fluidized Beds", *AIChE Symposium Series*, **69**, 115-122 (1973).

17. Yoshida H., K. Ito & N. Ise, "Microscopic Observation and Quasielastic Light-scatting Measurements of Colloid Crystals: Determination of the Radial Distribution Function and Structure Factor for the Two-state Structure" *J. Am. Chem. Soc.*, **112**, 592-596 (1990).

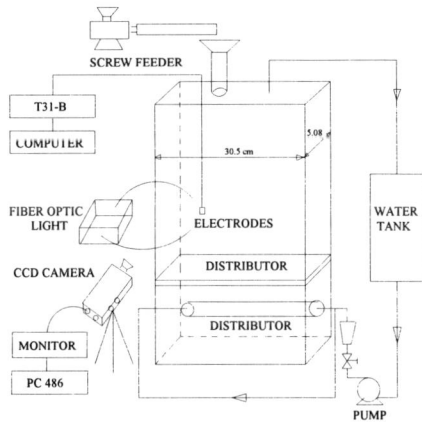

Fig. 1 Liquid-Solid Fluidized Bed With Instruments

Figure 2 A Typical Image Captured By CCD Camera For 50 μm Catalyst Particles

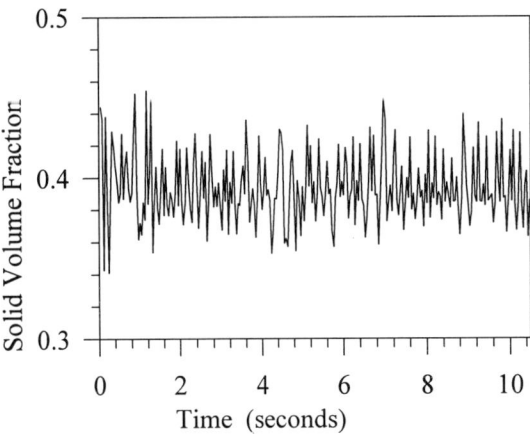

Figure 3 Instantaneous Solid Volume Fraction Measured By A Conductivity Probe

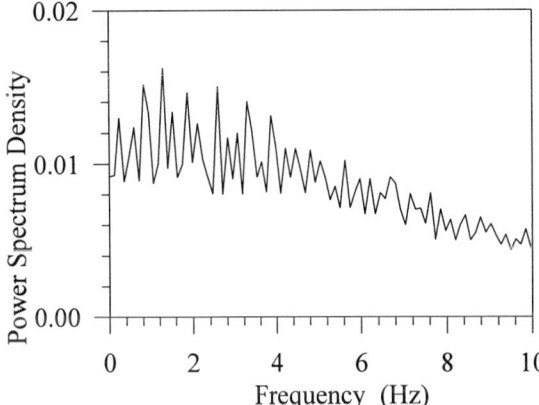

Figure 4 Power Spectrum Density of Solid Volume Fraction By FFT Method

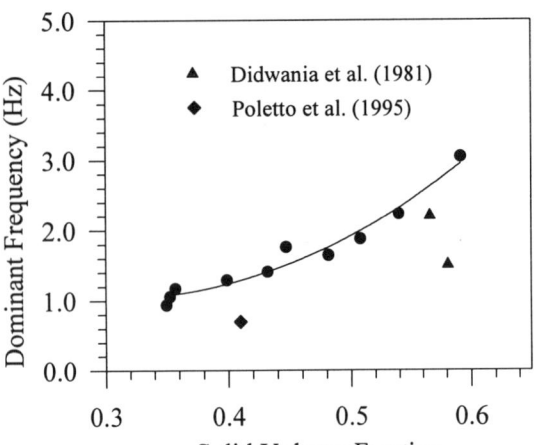

Figure 5 Dominant Frequency Profile For 50 μm Catalyst Particles in Liquid-Solid Fluidized Bed

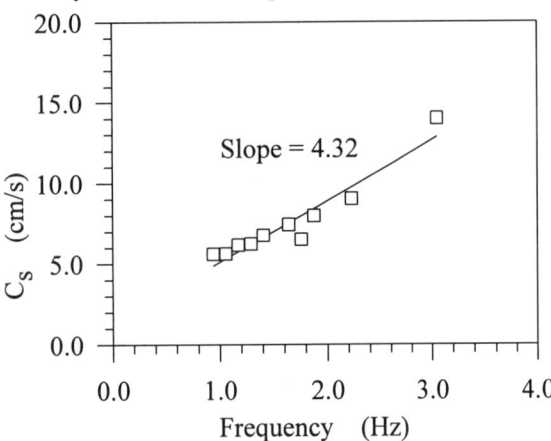

Figure 6 A Link Between The Dominant Frequency and The Particle Path Velocity Given by Particle Sonic Velocity

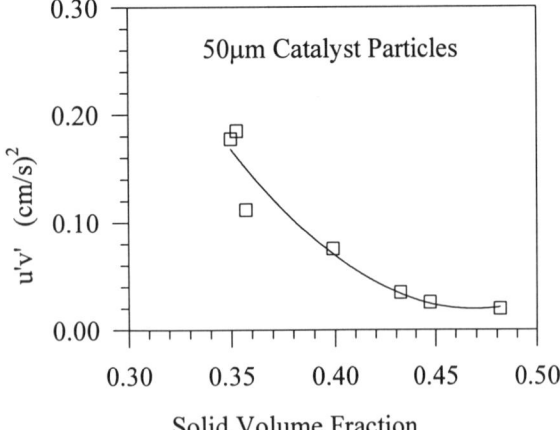

Figure 7 Variation of the Reynolds Stress with Solid Volume Fraction for 50 μm Catalyst particles

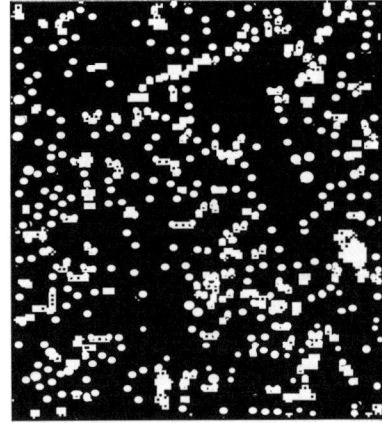

Figure 8 A Typical Particle Distribution Image

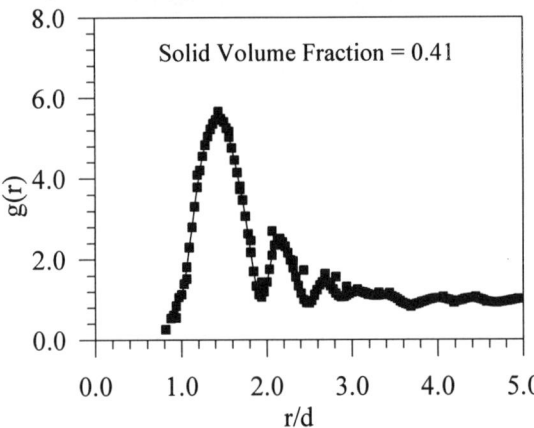

Figure 9 Radial Distribution Function Profile

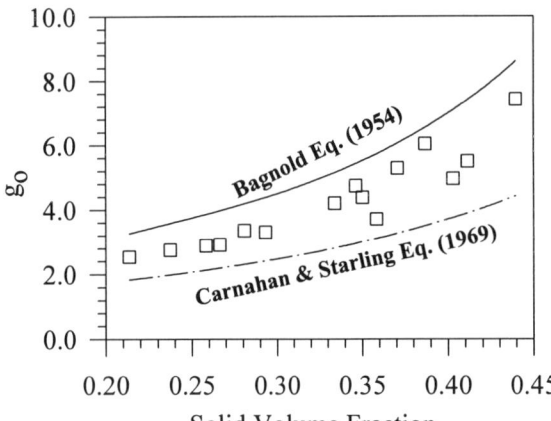

Figure 10 Comparison of Computed and Measured Radial Distribution Function for 450 μm Glass Particles In Liquid-Solid Fluidized Bed

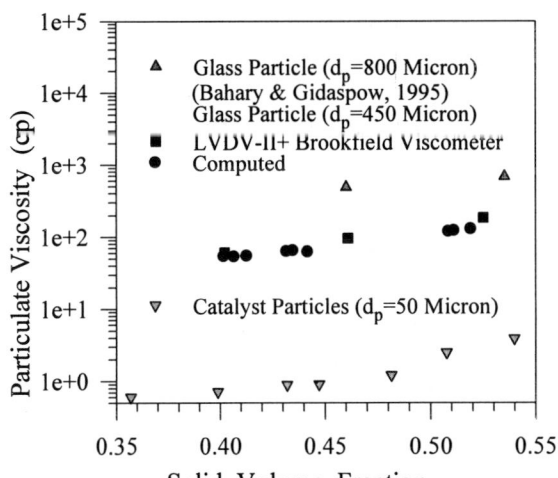

Figure 11 Comparison of Measured and Computed Particulate Viscosity In Liquid-Solid Fluidized Bed (For Dense Flow, Particulate Viscosity $\mu_s \propto d_p^2$)

Table 1 Summary of Experimental Data (50μm Catalyst Particles)

u_l cm/s	ε_s	u_p cm/s	σ_y cm/s	σ_x cm/s	θ cm/s^2
0.55	0.5917	0.4498	0.2118	0.2621	0.0528
0.77	0.5397	0.6199	0.2938	0.3891	0.1080
0.89	0.5076	0.7318	0.3770	0.5659	0.2015
1.11	0.4814	0.7851	0.3365	0.7632	0.2696
1.34	0.4471	0.9516	0.3438	0.8812	0.3376
1.54	0.4320	1.0490	0.4858	0.9315	0.4466
1.65	0.3989	1.0191	0.4811	1.1161	0.5695
1.76	0.3569	1.4564	0.5718	1.4239	0.8938
1.87	0.3525	1.4638	0.7581	1.4151	1.0506
1.98	0.3496	1.4814	0.8195	1.5111	1.2089

Table 2 Concept of Radial Distribution Function

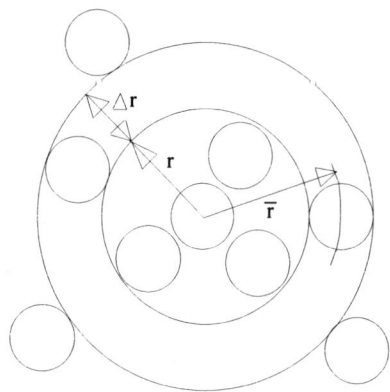

\# particles in shell between r and r+Δr:
$$\Delta N = 2\pi \bar{r}(\frac{N}{AREA})g(\bar{r})\Delta r$$
where: N is particles number in the observed area. Thus
$$\text{Local Density of Particles} = (\frac{N}{AREA})g(r)$$
where: $r = \sqrt{(x_i-x_j)^2 + (y_i-y_j)^2}$

To prevent two particles to be at the same location, for r≤d (particle size), g(r)=0. As r→∞, local density = local density × g(r). Hence, g(r)→1.0.

Table 3 Summary of Experimental Data (450μm Glass Beads)

u_l cm/s	ε_s	u_p cm/s	σ_y cm/s	σ_x cm/s	θ (cm/s)^2
2.47[1]	0.5188	1.775	5.857	2.176	14.592
2.47[2]	0.5108	1.821	6.234	2.881	18.488
2.47[3]	0.5083	2.298	6.389	2.702	18.475
3.28[1]	0.4415	2.146	6.187	3.151	19.378
3.28[2]	0.4346	2.578	6.469	3.833	23.746
3.28[3]	0.4305	2.857	6.981	3.406	23.977
4.47[1]	0.4126	2.631	6.657	3.866	24.736
4.47[2]	0.4064	2.983	6.941	3.926	26.337
4.47[3]	0.4015	3.233	7.831	3.671	29.466

1: measured position H=4.0 cm, 2: H=5.4 cm and 3: H=6.7 cm

Discontinuity in Particle Granular Temperature Across the Geldart B/A Boundary

George D. Cody and David J. Goldfarb
Exxon Corporate Research Laboratory, Clinton Township, Route 22 East, Annandale, NJ 08801

We present new values for the granular temperature or average particle kinetic energy of the dense phase of a fluidized bed consisting of monodispersed glass spheres, as a function of sphere diameter and gas flow. The non-intrusive experimental technique to measure the granular temperature, acoustic shot noise excitation of the wall of the bed by random particle impact, is first briefly described. The bulk modulus and other key parameters of the two phase fluid dynamics of the fluidized state, are then derived from the granular temperature, through a dense gas kinetic model.

We use these new results to explore the fundamental basis for the well known empirical distinction between fluidized particles which exhibit a single phase state at initial fluidization (Geldart A powders) and fluidized particles that exhibit gas bubbles at initial fluidization (Geldart B powders). Specifically we show that the experimental "jump" we observe in the granular temperature at the Geldart A/B transition is sufficient to account for the initial stability of the Geldart A phase on the basis of the one-dimensional, first-order, two-wave, stability theory first introduced by Jackson in the early sixties. We present new data on the diameter-dependent properties of the glass spheres during bed collapse and bed expansion, which demonstrate the distinction between Geldart A and B behavior for these monodispersed glass spheres. Finally, we present a simple Langevin model for the random velocity of the particle including inelastic collisions, to account for the dependence of the granular temperature on sphere diameter and gas flow, and discuss the implications of these new results for the fundamental physics of the Geldart A phase.

INTRODUCTION

We present the results of a new experimental investigation of the distinction between Geldart A particles, which exhibit a regime of stable homogeneous gas fluidization before bubbling, and Geldart B particles, which exhibit bubbles at fluidization[1]. Theoretical attempts to understand the stability of the gas fluidized state were begun in the early sixties, initially with simplified models, and more recently by direct numerical simulation of the equations of motion[2]. While instability is a broader category than bubbling, the Geldart B/A transition should be a feature of any first-principles theory of stability. Unfortunately, the absence of experimental data on key experimental quantities continues to be a significant obstacle to evaluating the relevance of stability theories to the Geldart classification[3]. The granular temperature, T^*, or mean squared particle fluctuation velocity, can be used, through the kinetic theory of dense gasses, to predict the bulk modulus, velocity of sound, and the Froude number of the dense phase of the fluidized state[4, 5] - all key parameters of current theories of stability and, presumably, the fundamental physics of the Geldart B/A transition. In this paper, we present new experimental data on the granular temperature, as a function of gas flow for monodispersed glass spheres whose diameters, D, span the Geldart B/A transition at D=120μm. We observe a discontinuity in the granular temperature across this transition. The magnitude of this discontinuity suggests that the initial stability of the Geldart A fluidized state for monodispersed glass spheres is a consequence of a bifurcation in the steady state particle dynamics between the two Geldart regions.

GRANULAR TEMPERATURE: DEFINITION

We denote the particle velocity as $\vec{c}(r,t)$ and its ensemble average defines the **particle drift velocity**, $\vec{V}(\vec{r},t)$, at the location "r", where $\vec{V}(\vec{r},t) \equiv <\vec{c}(r,t)>$. $\vec{V}(\vec{r},t)$ is often an obvious visual feature of a fluidized bed, for example, the observation of a downward "convective flow" of spheres at the wall. However, the sphere fluctuation velocity, $\vec{w}(r,t)=\vec{c}(r,t)-\vec{V}(\vec{r},t)$, is much too rapid to be a visual feature, and is consequently the major contribution to the granular temperature.

The granular temperature, T^*, is defined as the ensemble average of the squared fluctuation velocity, and is given by $3T^* \equiv <\vec{w}(r,t)^2> = <\vec{c}(r,t)^2> - [\vec{V}(\vec{r},t)]^2$. As noted, for fluidized beds $[\vec{V}(\vec{r},t)]^2 << <\vec{c}^2>$, and hence $3T^* \approx <\vec{c}(r,t)^2>$ With the critical assumptions of spatial uniformity and isotropy, T^* can then be expressed by one component of $<\vec{c}^2>$. For later convenience, we choose the velocity component normal to the wall, $v_n(r,t)$, and thus write $T^* \equiv (<\vec{w}(r,t)^2>/3) \approx <v_n(r,t)^2> \equiv v_n^2$.

If sphere collisions were elastic we would derive the granular pressure, bulk modulus, sound velocity, viscosity and diffusion constant of the particle phase from the kinetic theory of dense gasses and the velocity distribution function would be Maxwellian. Given inelastic collisions, such an approach is clearly only a first approximation[6]. Recent direct measurements of the normal stress of particle impacts on the wall in gas fluidized beds have been made by Campbell and collaborators[7] and Polashenski and Chen[8] as well as direct measurements of the dense phase viscosity by Gidaspow and Huilin[9]. The excellent agreement between these direct measurements of average dense phase parameters, and the same parameters derived from the dense phase kinetic model and our experimental data for the granular temperature [4, 5], encourages us to use this model to estimate fundamental parameters of the uniform fluidized state. Thus, we may define, with some confidence, a granular pressure P^*, granular bulk modulus E^* and granular sound velocity, C^* by the usual expressions from a dense kinetic gas model: $P^* = \rho_s T^*$, $E^* = dP^*/d\ln\rho_s = P^*$, $(C^*)^2 = (E^*/\rho_s) = T^*$. The quantity ρ_s is the density of the dense phase given by $\rho_s = \rho_o (1-e_s)$ where ρ_o is the mass density of the spheres and e_s is the voidage of the dense phase.

EXPERIMENTAL

The particles utilized in the present experiments are **monodispersed glass spheres** (1600 Series, Spacer Grade Microbeads, Cataphote Inc., Jackson, Miss., "90% true" [maximum variation of ± 7% from average diameter]) of average diameter 297, 210, 149, 105, 88, 74 and 63 microns (μm) and density ρ_o = 2.46 gm/cc. These spheres differ significantly in size, shape, and surface uniformity, from either sand or cat-cracking catalyst which in the past have often been the particles of choice for experimental studies of fluidization. The spheres span "Region B", and extend into "Region A" which for glass spheres starts at D=120μm.

Our measurements utilize **acoustic shot noise excitation** of the wall of the fluidized bed by random particle impact to determine the average granular temperature at the wall of the fluidized bed. Specifically we measure the wall acceleration power spectrum, $S_a(f,R_i)$, as a function of frequency f, at accelerometer location R_i=0. The technique and its validation have been exhaustively discussed in our previous publications[4, 5], and we will only summarize the defining equations. The **wall acceleration power spectrum**, excited by acoustic shot noise, is given by:

$$S_a(f,0) = <|H(f)|^2> <(2mv_n)^2 \rho_b v_n> A \quad (1)$$

In Eq. (1), "$<|H(f)|^2>$" is the frequency dependent, mean squared "transfer function", which is **experimentally obtained** by hammer excitation over the wall of the cylinder confining the fluidized bed; "m" is the mass of an individual sphere of density ρ_o; "v_n" is the mean squared particle fluctuation velocity at the wall; the volume density, "ρ_b", of spheres in the dense phase, is defined by $\rho_s = m\rho_b = (1-e_s)\rho_o$; "A" is the area of the wall of the cylinder.

The measured quantity is the mean-squared acceleration of the wall, a_m^2, in the *frequency range, 10-20kHz*, where for our experiments, acoustic shot noise has been shown to dominate other excitation sources. We define the quantity, a^2, as the integral of $S_a(f,0)$ over the frequency range 10-20kHz. In general, $a_m^2 = a^2 + a_n^2$ where a_n^2 is contributed by an electronic, or vibrational, noise power spectrum that is independent of gas flow. In our experiments the noise power spectrum is dominated by electronic noise. In what follows we define $a^2 \equiv [a_m^2 - a_n^2]$. We easily obtain from Eq. (1), the above definition of T^*, and the assumption of spatial uniformity and isotropy, the defining equation of the acoustic shot noise probe for the average T^* at the wall:

$$T^* \equiv v_n^2 = \left(\frac{a^2}{4I^2 A m \rho_s}\right)^{2/3} \quad (2)$$

where the quantity "I" is the integral of $<|H(f)|^2>$ over the range 10-20kHz, and all the other quantities have been previously defined.

GELDART POWDER CLASSIFICATION

Fig. 1a, locates the glass spheres of the present experiment on the "Geldart Plot" of powder fluidization behavior. The line separating Region A from Region B was originally determined by Geldart[10] through a comparison of two critical gas velocities: the empirical minimum gas superficial velocity for bubbling, U_{mb}, and that for fluidization, U_{mf}. Region A, "aeratable", was defined by $U_{mf} < U_{mb}$, and Region B, "bubbles readily" was defined by extrapolation and $U_{mf} > U_{mb}$.

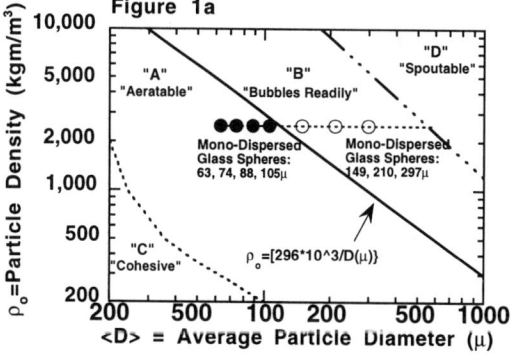

In Fig. 1b we exhibit our measurements of U_{mf} determined by the rapid onset of T^*, which agrees for the monodispersed spheres with the determination of U_{mf}, by either bed pressure drop or bed height. From Fig. 1b we note that the original empirical distinction between Geldart A and B for the glass spheres, becomes ambiguous since in region A $U_{mf} \approx U_{mb}$. However the factor of two increase in U_{mf} across the boundary implies a 15% increase in voidage at U_{mf} - suggestive of a higher granular temperature in Geldart A over B which is consistent with our direct measurements of T^*.

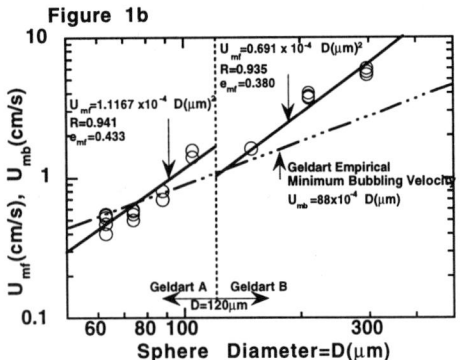

Figure 1b

CHANGE IN GRANULAR TEMPERATURE AT THE GELDART B/A TRANSITION

In Fig. 2a we show the RMS fluctuation velocity, v_n, scaled to the argon superficial velocity U_s, and particle diameter D, for Geldart B glass spheres with D=297, 210, 149µm, and for glass spheres with D=105µm, which are adjacent to the B/A boundary. For Geldart B glass spheres the data of Fig 2a introduces a *new scaling length for the granular temperature*, D_o, defined by $v_n = U_s(D_o/D)$. In Fig. 2a, we note a **factor of 1.5 increase** in D_o, (**D_o=187µm**), at U=(U_s/U_{mf})=2, for Geldart A glass spheres with D=105µm, compared to the Geldart B glass spheres with 149≤ D≤ 297µm (**D_o=121µm**)

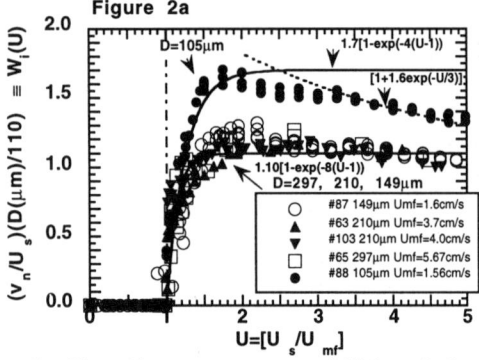

In Fig. 2b, we note an **additional factor of 1.5 increase** in D_o (**D_o=297µm**), at U=2 for Geldart A glass spheres with 63µm≤D≤88µm.

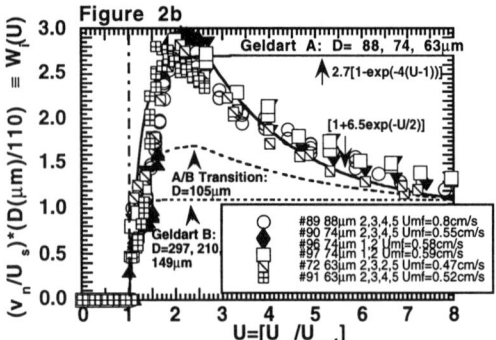

From the fitted curves of Figs. 2a and 2b, it is possible to construct $T^*(D,U)$ for the glass spheres in each of the Geldart regions. From the dense phase kinetic model, the granular pressure is then obtained from $T^*(D,U)$ as $P^*(D,U) = \rho_s T^*(D,U)$.

The derivative of $P^*(D,U)$ with respect to voidage, $E^s(D,U) = (-\partial P^*/\partial e_s) = \rho_o T^*(D,U)$, is the bulk modulus of the uniform fluidized state, a critical parameter of the first-order, one-dimensional model introduced by Jackson and later collaborators [11, 12, 13] more than 30 years ago. Before we discuss this model in detail, we show in Figs. 3a and 3b the quantity $E^s(D,U)$ as a function of U from the fitted curves of Figs. 2a and 2b, for D=105, 149, and 210µm in Fig. 3a and for D=63, 74, 88, and 149µm in Fig. 3b.

From Figs. 2a and 2b we note that for U ≥ 1.5, T^* and E^s are quadratic functions of D. Thus the comparable magnitude of E^s for D=105µm, and D=210µm in Fig. 3a is remarkable!

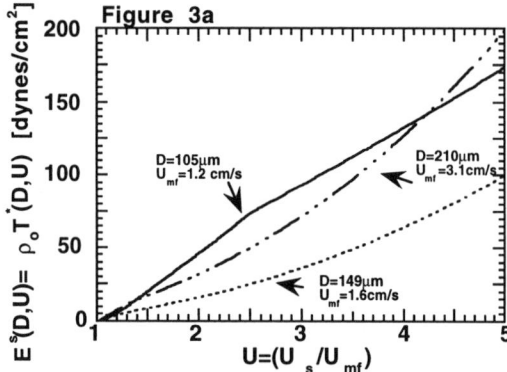

In Fig. 3b we compare Geldart A spheres with D=63, 74, 88µm with the Geldart B sphere with D=149µm. We note a dramatic change in both the magnitude of E^s and its dependence on U, for the Geldart A glass spheres.

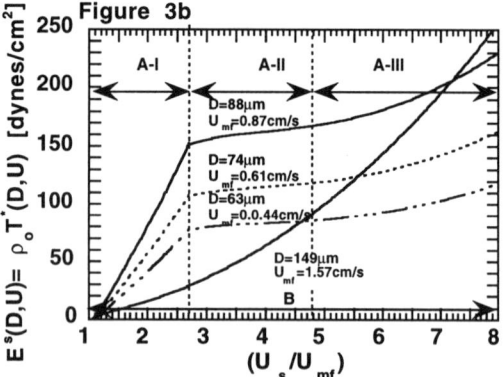

As shown in Fig. 3b, the **Geldart A glass spheres exhibit three distinct regions of fluidization**: **A-I**, where E^s increases quadratically with U, **A-II** where E^s is approximately constant, and

A-III where E^s again increases quadratically. It is tempting to identify region A-II with the "two phase" fluidization model of Davidson[14], where gas flow through the dense phase remains at minimum fluidization, and excess gas flows through bubbles.

STABILITY OF GAS FLUIDIZED BED

The simplest stability theory starts from a consistent set of equations that satisfy locally averaged momentum and mass balance for the gas and particles that comprise the fluid bed as well as the interactions between them[2, 3, 15] and from these develops the time-dependent equations governing one-dimensional, first-order perturbations from the uniformly stabilized state. Two critical velocities emerge that determine the stability of the fluidized state against small perturbations in voidage. The first critical velocity is the "**continuity velocity**", given by $U_e(D,U)=(1-e_s)(dU_s/de_s)$, which is the *velocity of a voidage wave* From the Richardson-Zaki equation, $U_s(e_s)=U_t e_s^n$, where U_t is the Stokes Velocity give by, $U_t=\rho_o g D^2/18\mu_g$, where g is the gravitational constant, and μ_g is the viscosity of the gas. We thus readily obtain for the "**continuity velocity**", $U_e(D,U) = U_t n e_s^{n-1}(1-e_s)$. The second critical velocity of the stability theory is the "**dynamic velocity**", $V_w(D,U) = (E^s/\rho_s)^{0.5}$, which is the *velocity of a pressure wave*.

In general, the uniform fluidized bed is stable against small perturbation in e_s, if $V_w(D,U) > U_e(D,U)$[16, 17]. In Fig. 4, we exhibit $V_w(D,U=2)$, as a function of D obtained from Figs. 2a and 2b, and $U_e(D,U=2)$ for argon with μ_g =210µpoise. From Fig. 1b we choose n = 4.70 for D≥149µm and n=4.85 for D≤105µm. From Fig. 4 we note that the dramatic increase in the normalized fluctuation velocity v_n shown in Fig. 2b, has just the right magnitude to make the dense phase of Geldart A glass spheres stable, at U=[U_s/U_{mf}]=2!

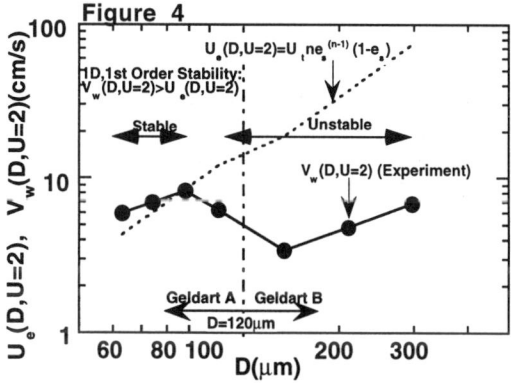

Figure 4

Previous attempts to obtain data for E^s utilized the bulk modulus of particles in mechanical contact held by suitable "cohesive forces". The early experimental papers of Rietema and collaborators[18] belong to this category, as well as recent theoretical work of Bouillard and Gidaspow [19]. These papers are important in focusing on the potential for particle contact in powders such as catalysts which have a broad, often log normal, distribution in particle size, shape and roughness and where there is, in addition, other experimental evidence for the role of contact forces at the fluidization transition [20]. However, it is unlikely that such particle "forces" play a significant role for the monodispersed glass spheres studied in the present paper.

Another approach has been taken by Foscolo, Gibilaro and collaborators [21, 22, 23, 24]. These authors develop a model for the velocity of the dynamic wave based on a "general theory of fluid-particle interaction". The velocity of the dynamic wave, V_w, in their model is given by $V_w=(3.2gD[1-e_s])^{0.5}$. Finally, Batchelor [25] arrived at an expression for V_w that is similar in form to Foscolo and Gibilaro. However, his approach focuses on the mean-squared particle fluctuation velocity obtained through an analysis of particle diffusion. As we pointed out earlier[4, 5] the average particle diffusion constant is proportional to the granular temperature $T^*(D,U)$, and $T^*(D,U)$ is thus a critical parameter of Batchelor's stability theory as well[3]. Indeed, one of the motivations for our initial research was to respond to Batchelor's observation that "it should be possible to make measurements of....the non-dimensional mean-square particle velocity fluctuation in a uniform fluidized bed...[but] there are few published data".

In Fig. 5 we compare our experimental results for $V_w(D,U=2)$ with the theoretical expression of Foscolo and Gibilaro also at U=2. We note comparable magnitude and functional dependence on sphere diameter, D, in the Geldart B regime for the two expressions for $V_w(D,U=2)$, a significant result given the experimental support for the Foscolo and Gibilaro expression [24]. From Fig. 5, we note that the Foscolo and Gibilaro value for $V_w(D,U=2)$ is too low by a factor of 1.2-2 to predict stability for 63µm≤D≤88µm, although their value for $V_w(D,U=1)$ does predict stability for glass spheres with D≤63µm.

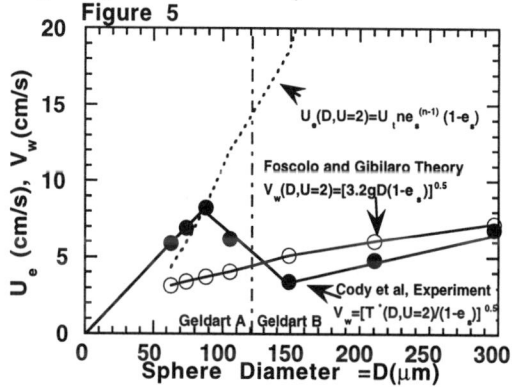

Figure 5

DISCUSSION

Experimental Differences Between Geldart A Powders and Geldart A Monodispersed Glass Spheres

The properties of Geldart A powders have been largely defined by particles such as catalysts which exhibit a wide dispersion in particle shape, surface and diameter, in sharp contrast to the monodispersed glass spheres that have been the focus of the present study. There are significant differences between the two types of particles that exhibit themselves in the simplest experiments. For example, we earlier noted the agreement for the monodispersed glass spheres between the minimum fluidization velocity determined by the onset of T*, and that determined by bed pressure drop or bed height. Such agreement does not hold for FCCU catalyst, where U_{mf} determined by T* onset is almost a factor of 3 larger than that determined by either bed pressure drop or bed height[5, 26].

Another significant difference between the two types of "Geldart A" particles, monodispersed glass spheres and log-normal dispersed FCCU catalysts, occurs in bed collapse measurements. In these experiments the bed height of a fluidized bed is followed in time after the shut-off of fluidization gas. The total collapse times for the Geldart A FCCU catalyst is about 20 times longer than that for the glass spheres as shown in Fig. 6 and discussed in more detail elsewhere[5].

Long bed collapse times have been associated with Geldart A behavior[27] and this dramatic difference raises questions on the **direct evidence** of the stabilization of the Geldart A phase for the monodispersed glass spheres. However in Fig. 6 we also compare bed collapse times for Geldart A and B glass spheres. and note a dramatic **increase** in the collapse time ΔT_c for Geldart A glass spheres compared to Geldart B glass spheres. The increase is reasonably fit by $\Delta T_c \approx D^{-3}$.

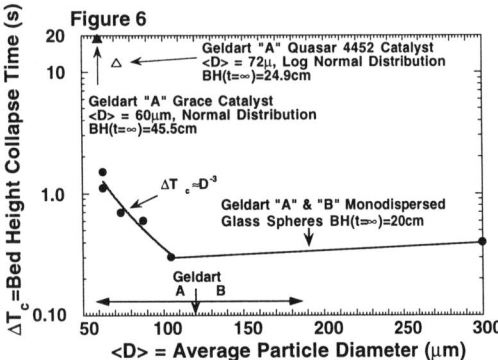

Bed expansion at fluidization is another experimental probe that exhibits a difference between Geldart A catalyst particles and glass spheres. For catalyst particles, bed expansion at the onset of bubbling is 2 to 3 times that exhibited by monodispersed glass spheres[26]. In Fig. 7 we show the normalized initial slope of the bed height, $\alpha = [\partial \ln(BH(U))/\partial U]$ at U=1 ($U_s = U_{mf}^{bh}$), for catalyst particles as well as glass and polymer spheres. For calibration with the expected magnitude of α, we show in Fig. 7, the value of $\alpha = (1/3n)$ derived from the Richardson-Zaki equation. Again we note a significant change in our data for α across the A/B transition for the glass spheres. Indeed, for Geldart A, $\alpha \approx D^{-2}$, and is larger than α for the catalyst!

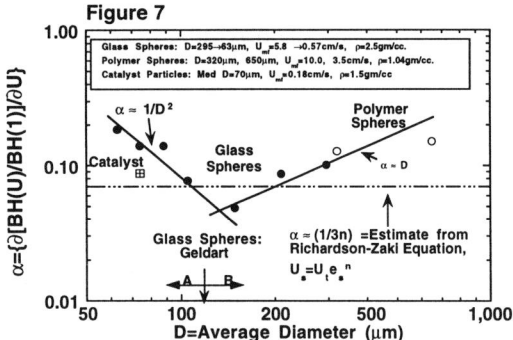

Scaling of T* With Diameter D, and Gas Flow U_s

Critical to all the numerical stability calculations is the magnitude and functional dependence of T* found in the present experiments. The quadratic dependence of T* on the gas superficial velocity, U_s was anticipated, the inverse quadratic dependence on sphere diameter was not! If we scale U_s with respect to U_{mf} ($U=U_s/U_{mf}$) and note that $U_{mf} \approx D^2$, we obtain $T^*(D,U) \approx U^2 D^2$. Is there any theoretical support for such a quadratic scaling of the granular temperature of the dense phase to the sphere diameter? What determines the new fundamental length scale, D_o, that is required to fit the data?

To address these issues we utilize a Langevin equation[28] for the random particle velocity $\vec{c}(r,t)$, which includes relaxation through inelastic collisions,

$$m\left(\frac{d\vec{c}(r,t)}{dt}\right) + m\left(\frac{\vec{c}(r,t)}{\tau_p}\right) = \vec{\mathbb{F}}(r,t) \qquad (3)$$

In Eq. (3). $\vec{\mathbb{F}}(r,t)$ is the random force exerted on the sphere by the fluidizing gas and impact of other particles, m is the mass of the sphere, and τ_p is the velocity relaxation time arising from inelastic collisions. Our earlier studies of the time dependence of the granular temperature during bed collapse gave $\tau_p \approx 90\text{-}150$ ms, a magnitude that has been shown to be quantitatively consistent with the expected coefficient of restitution for glass spheres of $e_p = 0.86\text{-}0.92$ [5]. In the steady state

$$T^*(D,U) = \frac{\langle \vec{c}(r,t) \cdot \vec{c}(r,t) \rangle}{3} = \left(\frac{\tau_p}{3m}\right) \langle \vec{\mathbb{F}}(r,t) \cdot \vec{c}(r,t) \rangle \quad (4)$$

The power input per unit volume $d\Omega^*/dt$ to maintain $T^*(D,U)$ constant is thus

$$d\Omega^*/dt = \rho_b \langle \vec{\mathbb{F}}(r,t) \cdot \vec{c}(r,t) \rangle = \left(\frac{3\rho_m T^*(D,U)}{\tau_p}\right) \quad (5)$$

This power loss is supplied by the fluidizing gas. The power per unit volume, $d\Omega^{fb}/dt$ to drive gas of viscosity, μ_g, through a "fixed bed" of voidage, e_s is readily derived from the Ergun equation[29] as

$$d\Omega^{fb}/dt = (150)\left(\frac{(1-e_s)^2}{e_s^3}\right)\mu_g\left(\frac{U_s^2}{D^2}\right) \quad (6)$$

If we assume that $d\Omega^*/dt = K \, d\Omega^{fb}/dt$ where K is a constant

$$T^*(D,U) = (50K)\left(\frac{(1-e_s)}{e_s^3}\right)\left(\frac{\mu_g\tau_p}{\rho_o}\right)\left(\frac{U_s^2}{D^2}\right) \quad (7)$$

and hence, we can define the scaling parameter D_o,

$$D_o^2 \equiv (50K)\left(\frac{(1-e_s)}{e_s^3}\right)\left(\frac{\mu_g\tau_p}{\rho_o}\right) \quad (8)$$

At $(U_s/Umf)=2$, $\mu_g=210\mu poise$, $\rho_o=2.5gm/cc$, $\tau_p=120ms$, with e_s from the Richardson-Zaki Equation, and the data of Figs. 1b, 2a, 2b, we obtain the following estimates for the constant K: for **Geldart B glass spheres, $D=297-149\mu m$., K=5%** ($D_o=121\mu m$, n=4.70, $e_s=0.439$); for **Geldart A/B glass spheres, $D=105\mu m$:, K=20%** ($D_o=187\mu m$, n=4.85, $e_s=0.497$); and for **Geldart A glass spheres, $D=88-63\mu m$:, K=50%** ($D_o=297\mu m$, n=4.85, $e_s=0.497$).

Increase in T*(D,U) Across the Geldart A/B Boundary: Cause or Effect?

The remarkable and systematic increase in $T^*(D,U)$ as the Geldart B/A boundary is crossed can be interpreted in two ways. The first would focus on the suppression of bubbles near U=1 for Geldart A glass spheres with the experimental consequence of increasing gas flow through the dense phase compared to Geldart B glass spheres. The rapid, essentially discontinuous increase in $T^*(D,U)$ found experimentally for $D\leq105\mu m$ would thus be experimental evidence of the sharpness of the Geldart B/A transition and, hence, additional validation of the acoustic shot noise as a quantitative non-intrusive probe of the particle granular temperature. Indeed, such an point of view was taken in our earlier publications[4, 5].

Another interpretation starts with the evidence of Figs. 2a,b which suggests a series of bifurcations in the dynamics of the glass spheres as the Geldart B/A boundary is approached and passed. *The Geldart B/A transition, and the unique properties of the Geldart A phase would then derive from the discontinuous increase in the granular temperature!* The fundamental physics for the apparent bifurcation ratio of 1.5 in v_n for $D=105\mu m$ relative to $D=149, 210, 297\mu m$, and the apparent bifurcation ratio of 1.5 in v_n for $D=63, 74, 88\mu m$ relative to $D=105\mu m$, remains a challenge for theory, as well as a note of caution for "experimental" simulations.

ACKNOWLEDGEMENTS

We are grateful to Roy Jackson and Morrel Cohen for helpful and clarifying discussions.

NOTATION

a^2 = Integral of $S_a(f,0)$ in frequency range 10-20kHz
a_n^2 = Integral of $S_n(f)$ in frequency range 10-20kHz
A = Internal area of cylinder confining fluid bed
$D = D(\mu m)$ = Particle diameter in microns (μm)
D_o = Scaling constant for v_n with dimensions of length in μm defined by $T^*(D,U_s)=U_s^2(D_o/D)^2$
e_s = Voidage of dense phase
E^* = Granular bulk modulus = $(dP^*/dln\rho_s)$
$E^s(D,U)$ = $(-\partial P^*/\partial e_s)$=bulk modulus of uniform fluidized state
$H(f,R_i)$ = experimental transfer function from force at R_i to accelerometer at $R_i=0$
$\langle|H(f)|^2\rangle$ = Experimental squared modulus of $H(f,R_i)$ averaged over the wall of the cylinder confining the fluidized bed
I = Integral of $\langle|H(f)|^2\rangle$ over the range 10-20kHz
K = Ratio of power required to maintain steady state granular temperature to viscous power input into fluid bed
m = Mass of glass sphere of diameter D.
n = Richardson-Zaki number defined by, $U_s = U_t e_s^n$
$S_a(f,0)$ = Acceleration Power Spectrum at frequency f and wall location $R_i=0$
$S_n(f)$ = Noise power spectrum that is independent of gas flow
$T^*(D,U)$ = Granular Temperature as function of sphere diameter and scaled gas superficial velocity
$U = (U_s/U_{mf})$
$U_e(D,U)$ = "Continuity Velocity" of 2-Wave Stability Theory = $-(1-e_s)(dU_s/de_s)$
U_{mb} = Gas superficial velocity at bubbling on-set(cm/s)
U_{mf} = Gas superficial velocity at fluidization on-set(cm/s)
U_{mf}^{bh} = Minimum fluidization velocity determined by bed expansion (cm/s)
U_s = Gas superficial velocity (cm/s)
U_t = Stokes velocity = $(\rho_o g D^2/18\mu_g)$
$\vec{V}(\vec{r},t) \equiv \langle\vec{c}(\vec{r},t)\rangle$ = Average particle drift velocity
$V_w(D,U)$ = "Dynamical Velocity" of 2-Wave Stability Theory = $(E^s/\rho_s)^{0.5}$
v_n^2 = Mean squared fluctuation velocity normal to wall = T^*
$\vec{c}(\vec{r},t)$ = Vector velocity of particle at location "$\vec{r}$" at time t.
$\vec{w}(\vec{r},t)=\vec{c}(\vec{r},t)-\vec{V}(\vec{r},t)$ = Particle fluctuation velocity at location "$\vec{r}$" at time t.

Greek Symbols

α = normalized slope of bed expansion at U=1
ΔT_c = Bed collapse time
ρ_o = Mass density of glass sphere
ρ_b = Volume density of spheres in dense phase
ρ_s = Mass density of dense phase =$\rho_o(1-e_s)$=$m\rho_b$
τ_p = Relaxation time for $\vec{c}(\vec{r},t)$ in Langevin Equation
μ_g = Viscosity of fluidizing gas

LITERATURE CITED

1. D. Geldart, in *Gas Fluidization Technology*, D. Geldart, Ed., John Wiley, New York, 1986, pp. 11-51.
2. R. Jackson, in *Fluid Particle Technology, AIChE Symposium Series 301*, A. W. Weimer, Ed., vol. 90, American Institute of Chemical Engineers, New York, 1994, pp. 1-30.
3. M. Nicolas, J.-M. Chomaz, E. Guazzelli, *Phys. Fluids*, **6**, (1994) 3936-3944.
4. G. D. Cody, D. J. Goldfarb, G. V. Storch, Jr., A. N. Norris, Particle Granular Temperature in Gas Fluidized Beds, AIChE Annual Meeting 11/12-17/95, Miami Beach, Florida (AIChE, 1995).
5. G. D. Cody, D. J. Goldfarb, G. V. Storch, Jr., A. N. Norris, *Powder Technology*, **87**, (1996) 211-232.
6. D. Gidaspow, *Multiphase Flow and Fludization - Continuum and Kinetic Theory Descriptions*, Academic Press, San Diego, 1994, p. 239-354.
7. C. S. Campbell, K. Rahman, *Meas. Sci. Tech.*, **3**, (1992) 709-712.
8. W. Polashenski, Jr., J. C. Chen, *to be published, Powder Technology*, , (1997)
9. D. Gidaspow, L. Huillin, Collisional Viscosity of FCC Particles in a CFB, AIChE Annual Meeting, 11/12-17/95, Miami Beach, Florida (AIChE, 1995).
10. D. Geldart, *Powder Technology*, **7**, (1973) 285-282.
11. R. Jackson, *Trans. Instn. Chem. Engrs.*, **41**, (1963) 13-21.
12. T. B. Anderson, R. Jackson, *I&EC Fundamentals*, **6**, (1967) 527-539.
13. T. B. Anderson, R. Jackson, *I&EC Fundamentals*, **7**, (1968) 12-21.
14. J. F. Davidson, in *Mobile Particulate Systems*, E. Guazzelli, L. Oger, Eds., Kluwer, Dodrecht, The Netherlands, 1995, pp. 173-220.
15. R. Jackson, in *Fluidization*, J. F. Davidson, R. Clift, D. Harrison, Eds., Academic Press, New York, 1985, pp. 47-72.
16. G. B. Wallis, *One-Dimensional Two-Phase Flow*, McGraw-Hill, New York, 1969, p. 122-242.
17. J. T. C. Liu, *Proc. R. Soc. Lond.*, **A389**, (1983) 331-347.
18. S. M. P. Mutsers, K. Rietema, *Powder Technology*, **18**, (1977) 239-248.
19. J. X. Bouillard, D. Gidaspow, *Powder Technology*, **68**, (1991) 13-22.
20. S. C. Tsinontides, R. Jackson, *J. Fluid Mech.*, **255**, (1993) 237-274.
21. P. U. Foscolo, L. G. Gibilaro, *Chem. Eng. Science*, **39**, (1984) 1667-1675.
22. P. U. Foscolo, L. G. Gibilaro, *Chem. Eng. Science*, **42**, (1987) 1489-1500.
23. P. U. Foscolo, L. G. Gibilaro, S. Rapagna, in *Developments in Fluidzation and Fluid Particle Systems, AIChE Symposium Series 308*, J. C. Chen, Ed., vol. 90, American Institute of Chemical Engineers, New York, 1995, pp. 44-50.
24. L. Gibilaro, P. Foscolo, R. di Felice, in *Two Phase Flow and Waves*, D. D. Joseph, D. G. Schaeffer, Eds., Springer-Verlag, New York, 1990, pp. 56-69.
25. G. K. Batchelor, *J. Fluid Mech.*, **193**, (1988) 75-110.
26. G. D. Cody, D. J. Goldfarb, Discontinuity in Particle Granular Temperature in Gas Fluidized Beds Across the Geldart B/A Boundary, M. Drake, J. Klafter, Eds., Symposium FF: Dynamics in Small Confining Systems-IV, Boston, Mass (MRS Fall Meeting, 1996).
27. J. R. Grace, in *Fluidized Processes, AICHE Symposium Series 289*, A. W. Weimer, Ed., vol. 88, American Institute of Chemical Engineers, New York, 1992, pp. 1-16.
28. R. J. Kubo, in *Reprint Series From Reports on Progress in Physics: Many Body Problems*, S. F. Edwards, Ed., Benjamin, New York, 1969, pp. 235-284.
29. D. Kunii, O. Levenspiel, *Fluidization Engineering*, Krieger, Malabar, Florida, USA, 1987, p. 73.

Flow Behavior in the Riser of a High-Density Circulating Fluidized Bed

A. S. Issangya, D. Rai, J. R. Grace and K. S. Lim
Department of Chemical Engineering, University of British Columbia
2216 Main Mall, Vancouver, Canada, V6T 1Z4

J. Zhu
Department of Chemical and Biochemical Engineering
University of Western Ontario, London, Canada, N6A 5B9

Voidage profiles and solids momentum were measured in the riser of a dual-loop circulating fluidized bed apparatus which was capable of providing high solids fluxes and cross-sectional average solids concentrations up to about 25% by volume over the entire height of the riser. For these high density conditions, the suspension is much denser near the wall than in the core of the riser, as in low density flows. However, no refluxing of solids was seen near the riser wall. Instead, the solids at the wall were found to be, on a time-average basis, moving slowly upwards.

There has been considerable work on the hydrodynamics of circulating fluidized beds. Most of the work reported in the literature has, however, been limited to CFB systems operating at relatively low solids suspension densities and solids circulation rates, conditions typical of CFB coal combustors. There is a major need to extend the studies to high density/high solids flux systems to improve understanding of the flow mechanics in existing catalytic processes, as well as to encourage development of reactors operating at even higher suspension densities and solid-to-gas feed ratios.

Knowledge of the solids distribution and flow behavior is key to successful design and operation of CFB systems. The solids distribution governs the pressure drop along the riser, residence time distribution of solids within the riser and heat transfer between the suspension and the wall. The cross-sectional average voidage along the height of a circulating fluidized bed riser commonly follows an S-shaped profile, with a dense region at the bottom, a dilute region at the top and an inflection point inbetween [e.g. [1] - [4]]. These profiles are influenced, among other factors, by gas and solids flowrates, solids entrance and exit configurations and gas distributor design [5]. Radially, there are strong concentration gradients [e.g. [4] - [8]] often represented by a core/annular model.

The above description applies to the upper regions of a typical dilute CFB riser. The few studies available on the lower dense region or in fully dense risers lead to different descriptions. Contractor et. al. [9] achieved solids holdups of nearly 0.2 over the entire height of a pilot scale CFB riser utilizing FCC particles and observed that there was little gas bypassing and that the predominant flow was not core/annular. Arena et. al. [10] found that as the solids circulation rate was increased, a "dense phase regime" of voidage between 0.75 and 0.85 moved up the riser; at $G_s = 190$ kg/m^{2}s and U = 3 m/s the dense phase filled the whole riser. In previous work [11], we observed that at high solids fluxes, the dense region having solids holdup between about 0.15 and 0.20 exhibited an almost homogeneous flow structure with negligible downflow of solids at the wall. Schnitzlein and Weinstein [12] found that air and solids flowrates had only a small effect on the dense region mean voidage; solids existed in at least three different forms: stagnant or downflowing near the wall with $\varepsilon \approx \varepsilon_{mf}$, travelling upward adjacent to the wall layer as dense aggregates, and dispersed dilutely in upward flowing gas in the core. Bai et al. [13] reported that, although solids moved both up and down in the dense phase region, the time-mean net flow direction was always upward. They also concluded that the bottom region had no clear core/annulus structure and was probably turbulent. Louge et al. [14] found that the average voidage across the riser in the dense phase was always greater than at the wall, indicating the presence of a denser annular region near the wall. While some studies [e.g. 15, 16] have considered the lower dense bed region to be in the turbulent flow regime, others [e.g. 17] have treated this region as a bubbling fluidized bed. Brereton and Grace [16] indicated that clustering predominates at the base of a riser, while core/annulus behavior is encountered towards the top.

EXPERIMENTAL APPARATUS

Our equipment is shown schematically in Fig. 1. The dual-loop circulating fluidized bed unit consists of two risers, two downcomers, a curved plate impingement separator, cyclones and a baghouse. All experiments were conducted in the first riser which has a diameter of 76 mm and a height of 6.10 m. The second riser of diameter 102 mm and height 7.62 m lifts the solids from the first downcomer to a higher level, facilitating a taller second downcomer. This design, coupled with a Roots blower of high capacity, gives sufficient solids inventory and pressure head to provide a wide range of solids circulation rates [18]. Both downcomers have internal diameters of 305 mm; they are 3.66 m and 7.62 m tall. Solids from the downcomers are fed into the risers via 76 mm gate valves which are aerated to avoid clogging.

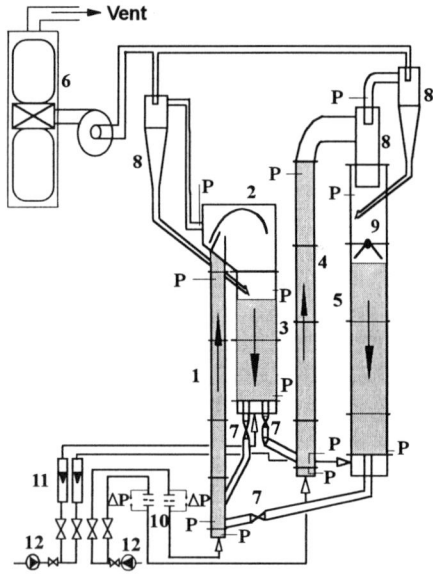

Fig. 1: Schematic diagram of high density CFB unit:
1. 1st Riser, 2. Impingement separator, 3. Storage tank, 4. 2nd riser, 5. Downcomer, 6. Baghouse, 7. Gate valve, 8. Cyclone, 9. Butterfly valve, 10. Orificemeter, 11. Rotameter, 12. Roots Blower, P Absolute pressure transducer, ΔP.Manometer.

FCC particles of mean diameter 70 μm and density 1600 kg/m^3 were used in the experiment. Their size distribution is given in Table 1. During each experiment, the air velocity was set in both risers, and both downcomers were brought to minimum fluidization before solids feed valves were opened. The system was deemed to have reached steady state when the solids levels in the downcomers remained unchanged. Further details are given elsewhere [11, 19].

A combined reflective optical fiber and momentum probe (Fig. 2) was developed to simultaneously measure local voidage and the net vertical momentum of the suspension. Fiber optic probes have been widely used to measure local voidages [e.g. 4, 8, 20] because of their simplicity, accuracy and low cost. Quartz fibers of diameter 0.015 mm carry light from a source and project it onto a swarm of particles. Interspersed fibers transmit light reflected by the particles to a phototransistor. The measuring surface was 2 x 2 mm square within a cylinder of 3 mm o.d. The probe was calibrated in a uniform downward stream of particles with two quick-closing slide plates to trap a small volume of the suspension surrounding the probe giving the solids concentration.

The Pitot type momentum probe consists of two brass tubes of 2.5 mm i.d. attached to the fibre optic probe and bent into smooth right angles to face in opposite directions, one up and the other down, as in the Cole pitometer and as in the designs of Azzi et al. [21] and Bai et al. [13, 22]. The distance between the tip ends is 15 mm. The pressure difference, ΔP_{mo}, was measured at ten radial positions. The pressure in each tube was measured by an absolute pressure transducer (Omega PX142) from which ΔP_{mo} was obtained by difference. Purge air is metered into both tubes using needle valves to prevent blockage. The purge air flowrate was found to have no significant influence on the pressure difference, provided it was less than the local air velocities anticipated in the riser.

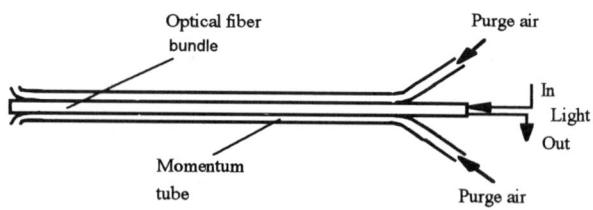

Fig. 2: Combined optical fiber momentum probe.

Table 1: Size distribution of FCC particles.

Mesh size [μm]	Mass fraction [%]
125 - 150	5.2
90 - 125	39.1
61 - 90	28.8
53 - 61	16.1
45 - 53	4.9
38 - 45	2.5
0 - 38	3.4

Since $\rho_p \gg \rho$, the gas contribution can be neglected. Neglecting also the static pressure difference between the two ports, we can write the following approximate equation, since the upstream facing port measures a stagnation pressure and the downstream facing probe measures the lesser pressure in its own wake:

$$\Delta P_{mo} \approx K \frac{1}{2} [\rho_p (1-\varepsilon) V_p |V_p|] \qquad (1)$$

K is dependent on the probe design, but is assumed to be independent of ε and V_p. The solids movement can then be characterized by recording ΔP_{mo} as a function of time. Since we are only interested in the solids flow direction in the current study, the calibration procedure as well as the solids velocity data are outside the scope of this paper. Signals from both the optical fibre probe and the absolute pressure transducers were logged on a personal computer via an A/D converter for periods of 100 s at a sampling frequency of 100 Hz. Differential pressure transducers were also connected to ports installed in the wall of the riser so that mean solids holdup profiles could be established. The solids circulation rate was measured by a drilled plate butterfly valve [11].

RESULTS AND DISCUSSION

Figure 3 shows axial solids hold-up profiles in the riser for U = 8 m/s and G_S up to 425 kg/m²s. The hold-ups were inferred from average differential pressures across sections of the riser by equating the static pressure drop to the bulk weight in the riser section. Neglecting wall friction or solids acceleration from the calculation has been shown [11] to have only a small effect in the region away from the inlet and does not significantly affect the findings of this work. As G_S is raised, the approximately exponential decay profile at the lowest circulation rate is replaced by sigmoidal profiles and then by more uniform profiles. Except for the bottom section, where relatively high apparent solids hold-ups result partly from particle acceleration, the solids hold-up remained 0.20 ± 0.05 over most of the height for 225 ≤ G_S ≤ 425 kg/m²s.

Radial profiles of local time-mean voidage at z = 1.57 m are plotted in Fig. 4 for U = 8 m/s. In all cases the riser was under dense phase conditions at this height. As for low G_S conditions, the concentration of solids is higher near the wall than in the core of the riser. However, whereas in dilute flows there exists a reasonably sharp transition from a denser annular layer to a dilute central region, the voidage transition becomes more gradual with increasing G_S. In addition, the voidage in the core is lower than in dilute systems. Increasing the solids circulation flux decreases the local voidage, the influence being greater near the wall than in the core.

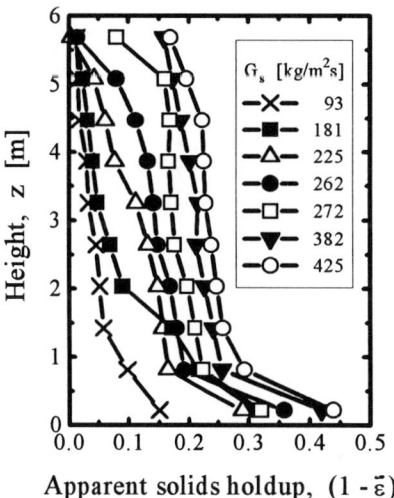

Fig. 3: Axial apparent solids holdup profiles at U = 8 m/s for various solids circulation rates.

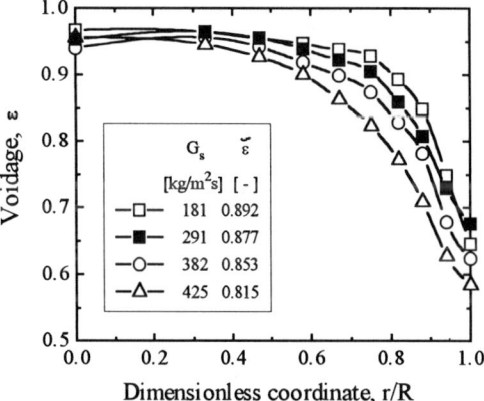

Fig. 4: Radial voidage profiles for U = 8 m/s, z = 1.57 m and various solids circulation rates.

Figure 5 plots the standard deviation of local voidage at different radial positions for the same conditions as in Fig. 4. The standard deviation increases from a minimum at the axis to a peak at r/R ≈ 0.9 before falling off toward the wall. The maxima in the voidage standard deviation profiles suggest vigorous solids interaction at the corresponding radial positions. A plot of the intermittency index [16] against radial position also indicated a peak between the wall and the center, approximately at the same location as the peak standard deviation [19]. The lower intermittency index values on either side of the peak suggest that the flow is more homogeneous in these areas.

Visual observations indicated that the dense phase adjacent to the wall has a relatively homogeneous structure, with no solids downflow at the wall for high G_s. To quantify the direction and relative magnitude of particle velocity, the momentum probe data are useful.

Figure 6 gives ΔP_{mo} traces at different radial positions 1.57 m above the distributor. At the center the flow is wholly upwards as reflected by the high positive value of ΔP_{mo}. Toward the wall the fluctuation amplitude decreases and some particles appear to move downwards, as reflected by brief periods of negative ΔP_{mo}. However, on average the net time mean flow is always upward as shown in Fig. 7 for G_S = 272 kg/m^{2}s and 354 kg/m^{2}s. For comparison purposes ΔP_{mo} for G_S = 126 kg/m^{2}s and z = 2.16 m, above the dense phase, is also plotted in Fig. 7. Here the flow is upward in the core region and downward at the wall. For the higher G_S values, the magnitude of the measured momentum probe pressure drop indicates that solids travel upwards rapidly in the central more dilute region and move upwards much more slowly at the wall.

Figure 8 gives local traces of instantaneous voidage at different radial positions together with corresponding probability distribution plots. A single high voidage peak is found at the center indicating that the flow is predominantly dilute (voidage > 0.9). Moving outwards, the suspension becomes denser and the voidage is distributed over a wider range. Near the wall the high voidage peak is no longer observed and is replaced by a high concentration peak at ε ≈ 0.6.

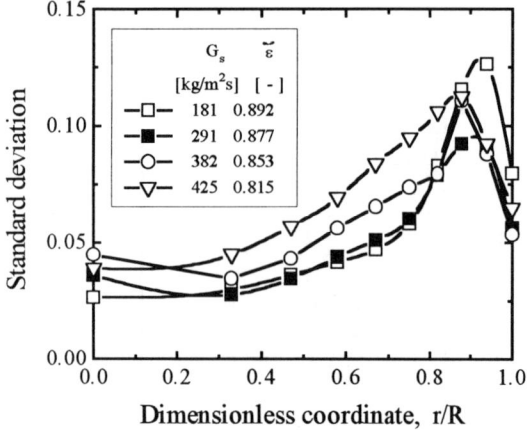

Fig. 5: Radial profiles of standard deviation of local voidage for the same conditions as in Fig. 4.

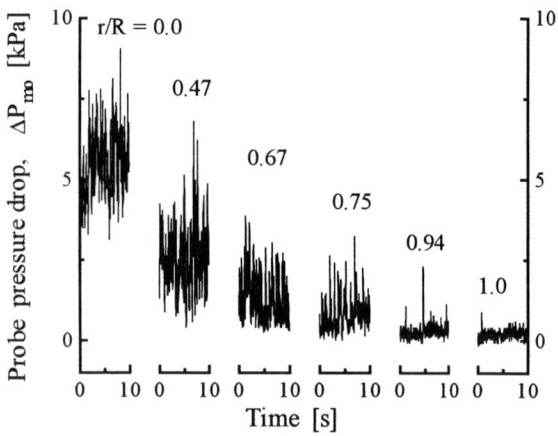

Fig. 6: Momentum probe pressure drop traces at six radial locations for U = 7.8 m/s, G_s = 354 kg/m^{2}s and z = 1.57 m.

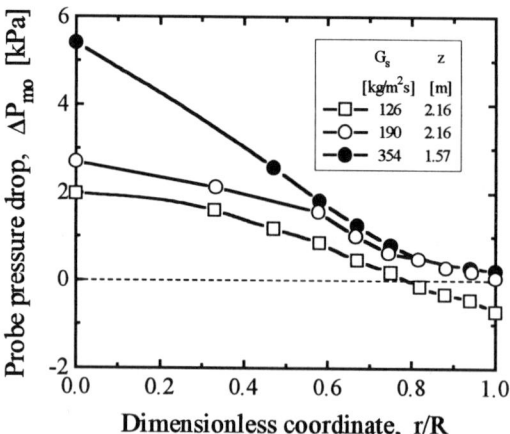

Fig. 7: Radial profiles of momentum probe pressure drop for U = 7.8 m/s.

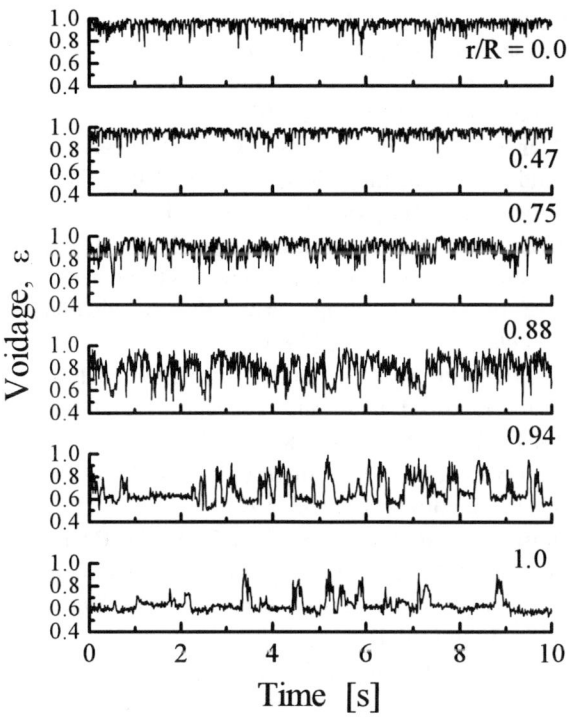

Fig. 8(a): Local voidage traces for six radial locations at U = 8 m/s, G_S = 291 kg/m²s and z = 1.57 m.

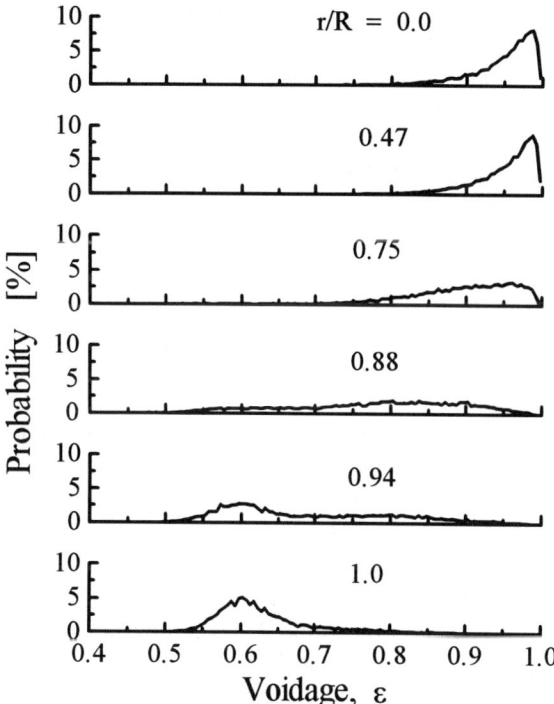

Fig. 8(b): Local voidage probability distribution plots for same conditions as in Fig. 8(a).

CONCLUSION

Strong radial particle concentration or voidage gradients observed in the top region of dilute CFB risers also exist in high density circulating fluidized beds. There is no region of net downflow at the wall. However, particles in the relatively dilute core still move upwards faster than those in the denser outer region. While there is no distinct core/annulus boundary, it may still be convenient to treat the suspension in terms of an inner dilute core and an outer more concentrated region, with the boundary between these two taken at the maximum standard deviation of local voidage fluctuations.

ACKNOWLEDGMENTS

Financial support from the Natural Sciences and Engineering Research Council of Canada is gratefully acknowledged.

NOTATIONS

G_S Solids circulation rate, kg/m²s
K   Momentum probe constant, -
r   Radial coordinate, m
R   Riser inner radius, m
U   Superficial gas velocity, m/s
V_p Particle velocity, m/s
z   Height above distributor, m
ΔP_{mo} Momentum probe pressure drop, Pa
ε Time mean local voidage, -
$\bar{\varepsilon}$ Cross-sectional average voidage, -
ε_{mf} Voidage at minimum fluidization, -
ρ Air density, kg/m³
ρ_p Particle density, kg/m³

LITERATURE CITED

1. **Li, Y. and M. Kwauk**, The dynamics of fast fluidization, in *Fluidization*, J. R. Grace and J. M. Matsen, eds., Plenum, New York, 537 - 544 (1980).
2. **Arena, U., A. Cammarota, L. Pistone and P. V. Tecchio**, The high velocity fluidization behavior of solids in a laboratory scale circulating bed, in *Circulating Fluidized Bed Technology*, P. Basu, ed., Pergamon, New York, 119 - 125 (1986).

3. Hartge, E.-U., Y. Li and J. Werther, Analysis of the local structure of the two phase flow in a fast fluidized bed, in *Circulating Fluidized Bed Technology II*, P. Basu and J. F. Large, eds., Pergamon, Oxford, 153 - 160 (1986).

4. Hartge, E.-U., D. Rensner and J. Werther, Solids concentration and velocity patterns in circulating fluidized beds, in *Circulating Fluidized Bed Technology*, P. Basu, ed., Pergamon, New York, 165 - 180 (1988).

5. Grace, J. R., Riser geometry influence on CFB particle and fluid dynamics, in *Circulating Fluidized Bed Technology V*, J. Li and M. Kwauk, eds. Science Press, Beijing, (1996).

6. Weinstein, H., M. Shao and M. Schnitzlein, Radial variation in solid density in high velocity fluidization, in *Circulating Fluidized Bed Technology*, P. Basu, ed., Pergamon, New York, 201 - 206 (1986).

7. Brereton, C. M. H., "Ph.D. dissertation", Univ. of British Columbia, (1987).

8. Horio, M., K. Morishita, O. Tachibana and N. Murata, Solid distribution and movement in circulating fluidized beds, in *Circulating Fluidized Bed Technology II*, P. Basu and P. Large, eds., Pergamon, Oxford, 147 - 153 (1988).

9. Contractor, R. M., G. S. Patience, D. I. Garnett, H. S. Horowitz, G. M. Sisler and H. E. Bergna, A new process for n-butane oxidation to maleic anhydride using a circulating fluidized bed reactor, in *Circulating Fluidized Bed Technology IV*, A. A. Avidan, ed., 466 - 471 (1994).

10. Arena, U., A. Cammarota, L. Massimilla and D. Pirozzi, The hydrodynamic behavior of two circulating fluidized bed units of different sizes, in *Circulating Fluidized Bed Technology II*, P. Basu and J. F. Large, eds., Pergamon, Oxford, 223 - 230 (1988).

11. Issangya, A. S., D. Bai, H. T. Bi, K. S. Lim, J. Zhu and J. R. Grace, Axial solids hold-up profiles in a high-density circulating fluidized bed riser, in *Circulating Fluidized Bed Technology V*, J. Li and M. Kwauk, eds., Science Press, Beijing (1996).

12. Schnitzlein, M. G. and H. Weinstein, Flow characterization in high-velocity fluidized beds using pressure fluctuations, *Chem. Eng. Sci.*, **43**, 2605 - 2614 (1988).

13. Bai, D, E. Shibuya, Y. Masuda, K. Nishio, N. Nakagawa and K. Kato, Distinction between upward and downward flows in circulating fluidized beds, *Powd. Tech.*, **84**, 75 - 81 (1995).

14. Louge, M., D. J. Lischer and H. Chang, Measurement of voidage near the wall of a circulating fluidized bed riser, *Powd. Tech.*, **62**, 269 - 276 (1990).

15. Bolton, L. W. and J. F. Davidson, Recirculation of particles in fast fluidized risers, in *Circulating Fluidized Bed Technology II*, P. Basu and J. F. Large, eds., Pergamon, Oxford, 139 - 152 (1988).

16. Brereton, C. M. H. and J. R. Grace, Micro-structural aspects of the behavior of circulating fluidized beds, *Chem. Eng. Sci.*, **48**, 2565 - 2572, (1993).

17. Svensson, A., F. Johnsson and B. Leckner, Fluidization regimes in non-slugging fluidized beds: the influence of pressure drop across the air distributor, *Powd. Tech.*, **86**, 299 - 312 (1996).

18. Bi, H. and J. Zhu, Static instability analysis of circulating fluidized beds and concept of high-density risers, *AIChE J.*, **39**, 1272 - 1280 (1993).

19. Issangya, A. S., "Ph.D. dissertation", Univ. of British Columbia, Vancouver, Canada, 1997.

20. Zhou, J., J. R. Grace, S. Qin, C. M. H. Brereton, C. J. Lim and J. Zhu, Voidage profiles in a circulating fluidized bed of square cross-section, *Chem. Eng. Sci.*, **49**, 3217 - 3226 (1994).

21. Azzi, M., P. Tulier, J. F. Large and J. R. Bernard, Use of a momentum probe and gammadensitometry to study local properties of fast fluidized beds, in *Circulating Fluidized Bed Technology III*, P. Basu, M. Horio and M. Hasatani, eds., Pergamon, Oxford, 189 - 194 (1991).

22. Bai, D., E. Shibuya, Y. Masuda, N. Nakagawa and K. Kato, Flow structure in a fast fluidized bed, *Chem. Eng. Sci.*, **51**, 957 - 966 (1996).

Solids Loading vs. Solids Flux as an Independent Parameter in CFB Similitude Studies

Ethan Brue[1] and Robert C. Brown[2]
Iowa State University, Ames, IA 50011

Research in our laboratory suggests that solids loading is a better choice than solids flux as an independent parameter in similitude studies of circulating fluidized beds. In addition, the coefficient of restitution for particle-wall collisions is an important hydrodynamic parameter.

Similitude analysis offers an important tool for the design of fluidized beds. Recent research has identified the following dimensionless parameters as important to hydrodynamics of circulating fluidized beds [1]:

$$[U^2/gD, \rho_s/\rho_g, \rho_g UD/\mu_g, H/D, G_s/\rho_s U, \phi, PSD] \quad (1)$$

The first term is the Froude number, Fr, the second term is the ratio of particle density to gas density in the fluidized bed, the third term is the hydraulic Reynolds number based on the diameter of the fluidized bed, Re_H, the fourth term is the ratio of bed height to bed diameter, the fifth term is the dimensionless solids flux through the reactor, and the sixth and seventh terms are the sphericity and size distribution of the particles in the reactors. Circulating fluidized beds that employ L-valves or other solids flow devices that depend on pressure balance between the riser and downcomer of the reactor should also include the total mass of particles, M_t, charged to the reactor. In dimensionless form, this last parameter is $M_t/\rho_s D^3$.

---

[1]Iowa State University, Department of Mechanical Engineering, Ames, Iowa. Currently with Black & Veatch Overland Park, Kansas

[2]Iowa State University, Departments of Mechanical Engineering and Chemical Engineering, Ames, Iowa

Researchers, although generally accepting the parameters identified by Glicksman as correct, have noted disturbing discrepancies between model and prototype reactors operating under presumed hydrodynamic similitude [2-5]. We have observed similar discrepancies in our own studies of circulating fluidized beds that we think arise from treating dimensionless solids flux as an independent parameter of the hydrodynamic state of a circulating fluidized bed [6]. The goal of this research is to demonstrate that solids loading is a better dimensionless parameter to describe solids circulation in the riser of a circulating fluidized bed.

EXPERIMENTAL METHOD

We conducted experiments in a 5 cm dia. model and a 10 cm dia. prototype. The model reactor is shown schematically in Figure 1; the prototype reactor is geometrically similar but twice as large as the model reactor. Time-series of pressure fluctuations was Fourier transformed and presented as Bode plots. Details can be found in Reference [6].

RESULTS AND DISCUSSION

The first experiment, performed in the 5 cm dia. reactor with 0.15 mm dia. steel shot for bed material, investigated the importance of total mass of solids in a circulating fluidized bed to its hydrodynamic behavior. We found that as the amount of solids charged to the reactor increased, the voidage at any given elevation in the riser of the reactor decreased - the bed became denser. As a result of this finding, subsequent evaluations of dynamic scaling were performed with dimensionless total solids loading, $M_t/\rho_s D^3$, set equal between prototype and model reactors to avoid this influence.

The second set of experiments investigated whether the scaling parameters given in Equation 1, along with $M_t/\rho_s D^3$, are able to predict hydrodynamic similitude between model and prototype beds. In general, comparisons between axial voidage profiles and Bode plots for prototype and model beds under these conditions were not favorable. An example of such a comparison is shown in Figures 2 and 3.

The next set of tests substituted solids loading in the riser, M_r, in place of solids flux as a dimensionless parameter. Solids loading can be monitored by noting the level of solids stockpiled in the riser and comparing this volume of solids to the total volume of solids charged to reactor. The appropriate dimensionless parameter for solids in the riser, $M_r/\rho_s D^3$, is equivalent to L_r/D, where L_r is the height of solids stockpiled in the downcomer and D is the diameter of the reactor.

Figures 4 and 5 show the results of similitude experiments conducted in the model and prototype CFB using the above parameters. The good match between axial voidage profiles and Bode plots suggests that the reactors are operating under near similitude conditions. However, the dimensionless solids fluxes were not well matched under these circumstances. The dimensionless solids flux in the model reactor was 55% higher than in the prototype reactor, which is well beyond the uncertainty of 15 - 20 % for solids flux measurements in our systems.

This result suggests that something is occurring within the prototype reactor that prevents particles from leaving the riser as readily as occurs in the model reactor. We hypothesized that the coefficient of restitution for collisions of glass beads with the Plexiglas top-plate in the large reactor is significantly higher than that for steel shot colliding with the aluminum top-plate in the small reactor. As a consequence, glass beads in the prototype reactor collide more energetically with the top of the riser and rebound back down into the riser as an internal recirculating flow whereas the momentum of the steel shot is dissipated upon collision with the top plate and are carried out of the riser with the gas flow. We tested this hypothesis by replacing the top plate of each riser with "dead-space" extensions, an example of which is illustrated in Figure 1. These extensions allowed the upward moving particles to change direction without impacting the end plate. Remarkably close matches in axial voidage profiles and Bode plots for model and prototype reactors were achieved in this test, as shown in Figures 6 and 7.

CONCLUSIONS

Solids loading, rather than solids flux, should be treated as an independent parameter to describe hydrodynamics in circulating fluidized beds. Furthermore, poor agreement between dimensionless solids fluxes for the reactors under these circumstances indicated that additional forces must be considered to successfully scale circulating fluidized bed reactors. Additional experiments illustrated the importance of including the effect of collisions between particles and the top plate of the riser if solids flux is to be successfully predicted. These experiments demonstrate that coefficient of restitution for particle-wall collisions is an important hydrodynamic parameter.

ACKNOWLEDGEMENTS

We appreciate the support of the U.S. Department of Energy for this work under contract DE-FG22-94PC 94210.

NOTATION

D   fluid bed diameter (m)
g   gravitational constant (m/s^2)
G_s solids flux (kg/m^2-s)
H   fluid bed height (m)
PSD   particle size distribution
U   superficial gas velocity (m/s)

μ_g gas viscosity (N-s/m^2)
ρ_g gas density (kg/m^3)
ρ_s particle density (kg/m^3)
ϕ particle sphericity

REFERENCES

1. **Glicksman L.R., M. R. Hyre, and P. A. Farrel.** "Dynamic similarity in fluidization," *Int. J. Multiphase Flow* 1993, 20, Suppl., 381-386, 1994.
2. **Martin-Letellier, S. and M. Louge.** "The role of gas density in circulating fluidized bed risers," Fluidization and Fluid-Particle Systems, H. Arastoopour, Ed.; American Society of Chemical Engineers: New York, 1995; 122-127.
3. **Glicksman, L.R., D. Westphalen, C. Bereton, and J. Grace.** "Verification of the scaling laws for circulating fluidized beds." In *Circulating Fluidized Bed Technology III*. P. Basu, M. Hasatani, and M. Horio, Eds.; Pergamon Press: Oxford, 1991; 199-124.
4. **Glicksman, L. R., M.R. Hyre, D. Westphalen,** "Verification of scaling relations for circulating fluidized beds," Proc. 12th Int. Conf. on Fluidized Bed Combustion, 1993, pp. 69-80.
5. **Westphalen, D. and Glicksman, L. R.** "Experimental verification of scaling for a commercial-size CFB combustor," in Circulating Fluidized Beds IV. A. Avidan, Ed. 1993, 436-441.
6. **Brue E. and R.C. Brown** "Validation of hydrodynamic similitude in fluidized beds via pressure fluctuations." In *Proc. 5th World Congress of Chemical Engineering*. AIChE, San Diego, July 1996.

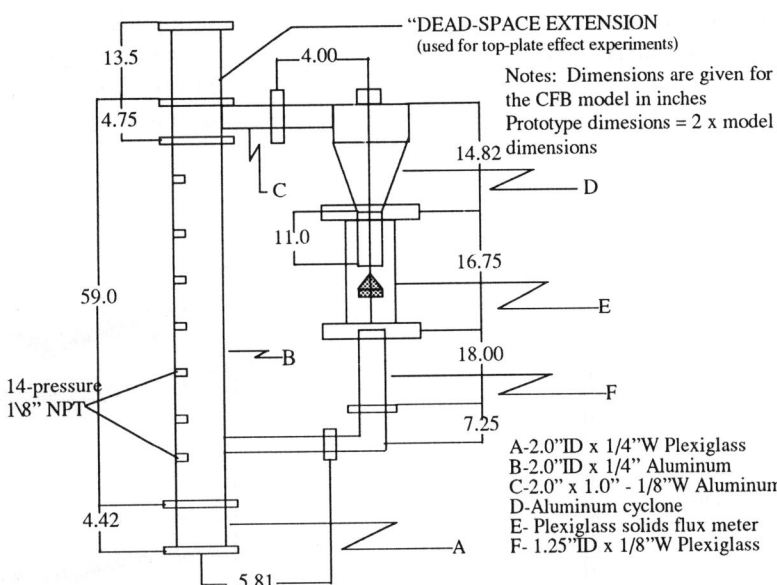

Figure. 1. Schematic of the model circulating fluidized bed (prototype fluidized bed is scaled twice the size of the model fluidized bed)

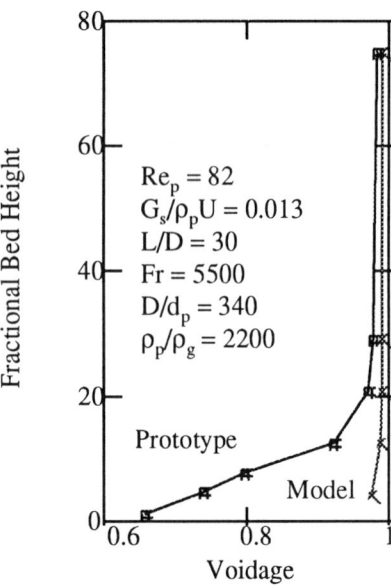

Figure 2. Axial profiles for model and prototype using solids flux as an independent parameter

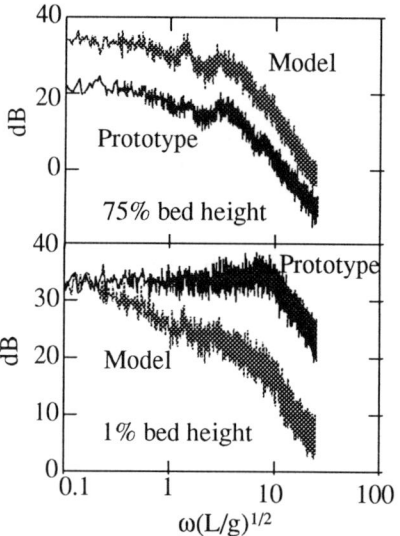

Figure 3. Bode plots of pressure fluctuations for model and prototype at two different bed elevations using solids flux as an independent parameter (same operating conditions as Figure 2)

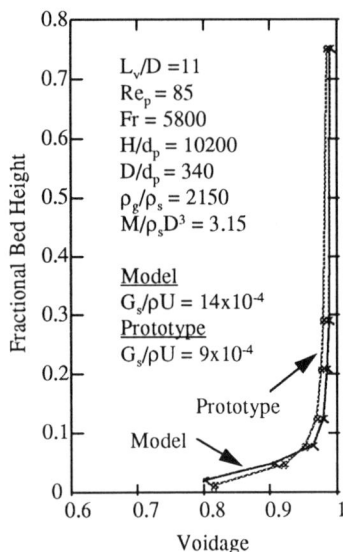

Figure 4. Axial voidage profile for model and prototype using solids loading in the riser as an independent parameter

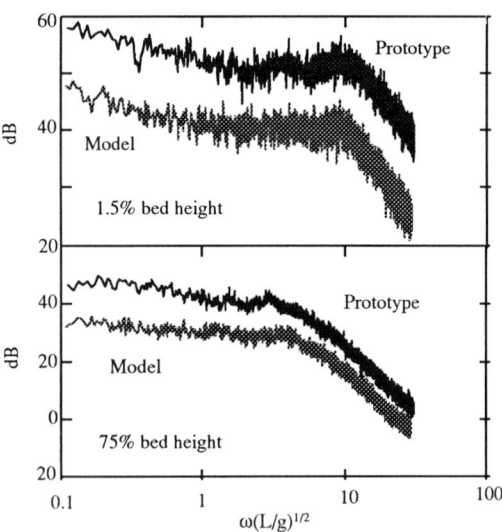

Figure 5. Bode plots of pressure fluctuations at two different bed elevations for model and prototype using solids loading in the riser as an independent parameter (same operating conditions as Figure 4)

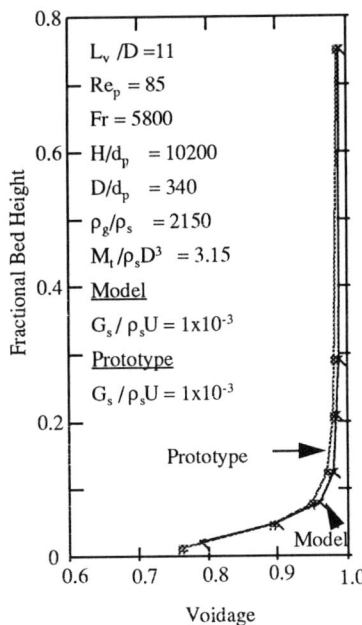

Figure 6. Axial voidage profile for model and prototype with dead-space extension using solids loading in the riser as an independent parameter

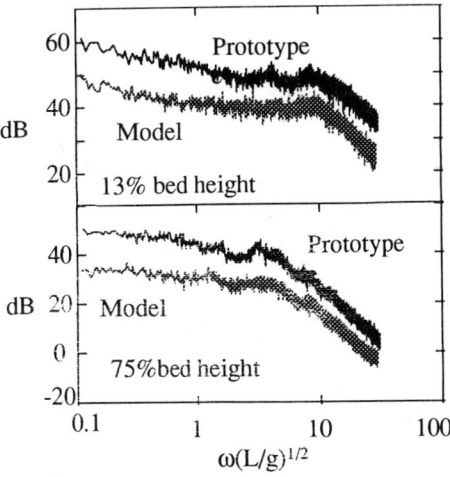

Figure 7. Bode plots of pressure fluctuations at two different bed elevations for model and prototype with dead-space extensions using solids loading in the riser as an independent parameter (same operating conditions as Figure 6)

A Detailed Experimental Description of the Flow in the Entrance Section of a FCCU

Apostolos E. Kostazos and Herbert Weinstein
CUNY Graduate School and Department of Chemical Engineering
The City College of New York, New York, NY 10031

An experimental study was carried out to examine the effect of the nozzle design on the gas-solid flow in the entrance region of a fluid catalytic cracking unit (FCCU). A mockup of a FCCU entrance section which was installed in the City College fast fluidization unit added two operating parameters, the radial gas injection position and the ratio between the primary and secondary of "Fluffing" air flowrates.

The results show clearly that the radial position of the gas injection plays the dominant role in determining the rapidity of the mixing of gas and catalyst. Gas injection at the center of the riser creates a flow dilute in catalyst near the center and dense near the wall. Gas injection near the riser wall, however, creates a generally flat distribution of catalyst across the entire riser.

INTRODUCTION

The entrance section of a fluid catalytic cracking unit (FCCU) riser consists of three major components[1]. These are a catalyst upflow section through which the catalyst is carried by "fluffing" steam, an area expansion section in which the oil nozzles are located, and the initial region of the straightside riser in which most of the oil vaporization and cracking takes place. In an attempt to study the influence of the oil nozzle design on unit performance, a mockup of a FCCU riser entrance section was installed in the City College fluidization facility[2].

In this study, the effect of the "fluffing" steam flow rate and the oil nozzle position on the gas-solid flow characteristics at the initial section 0-20 L/D of the riser are examined. The gas~solid flow in this initial section is critical for reactor performance since FCCU risers have a length usually of no more than 30 diameters. This makes the entire reactor an entrance region. This work includes measurements of the axial and radial solid distributions and the radial distributions of the axial solid flux and gas and solid estimated axial velocities. These data can give important information concerning the effect of the nozzle design on the riser performance.

EXPERIMENTAL

Figure 1 shows the new entrance section which consists of a contraction and an expansion section and four gas injection nozzles of 2.54cm ID symmetrically located above the expansion section. The nozzles can be moved to any radial position, while holding their axes parallel to the axis of the riser. The U-bend section is equipped with five nozzles and the air entering the system from these nozzles is considered as "fluffing" air and represents the "fluffing" steam of a FCCU riser.

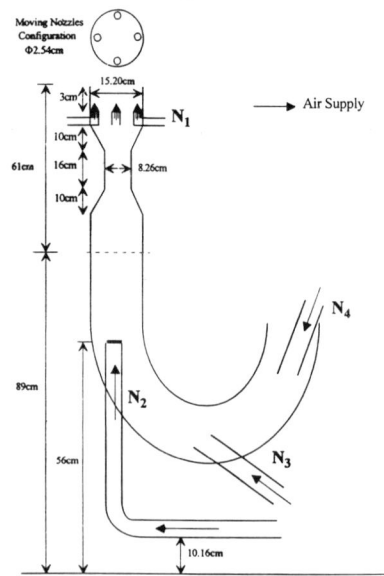

Figure 1 : Nozzle Configuration.

With this configuration two more operating parameters have been added, the ratio between primary and secondary or "fluffing" air flowrate and the position of the four nozzles.

RESULTS AND DISCUSSION

Figure 2 shows the axial apparent voidage profile for a superficial gas velocity U_g=3.8m/s and a solid circulation rate G_s=111kg/m²s. The nozzles are located as close as possible to the riser axis touching each other. Two different "fluffing" air flowrates are examined, 22% and 38% of the total flowrate. It can be seen that there is quite different behavior in the section from 0 to 1.50m above the nozzle inlets. For the low "fluffing" air case (FA=22%), the apparent void fraction increases from 0 to 0.4m followed by a section of complex flow up to 1.50m where the apparent void fraction increases and then decreases locally. When the "fluffing" air flowrate is higher (FA=38%), the apparent void fraction from 0 to 1.50m decreases smoothly from about 0.84 at 0m to about 0.80 at 1.50m. This behavior shows that the "fluffing" air flowrate effect is significant over this short section from 0 to 1.50m (10 L/D) when the nozzles are located as close as possible to the riser axis. Above the elevation of 1.50m the effect of the "fluffing" air flowrate is minimal. Figure 3 shows the effect of the "fluffing" air on the axial voidage profile when the nozzles are located as close as possible to the riser wall. It is observed that the "fluffing" air flowrate does not affect significantly the axial apparent voidage profile even over the initial section from 0 to 1.50m. The axial voidage profile for the case with the nozzles at the wall has an inflection point much higher than has the profile for the nozzles at the center case. Generally the profile when the nozzles are located at the wall is smooth without strong local changes of the apparent void fraction.

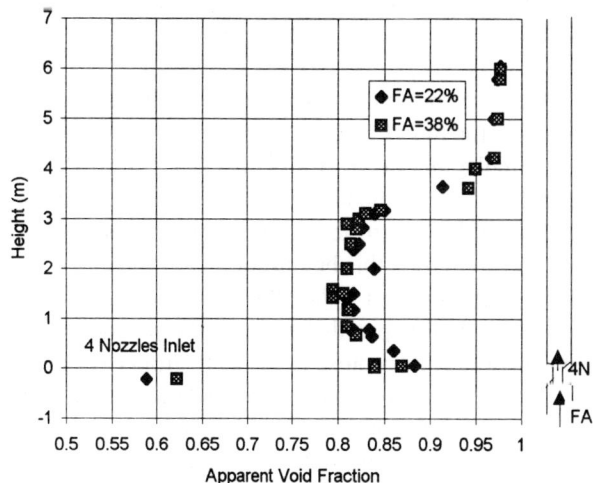

Figure 3: Axial Voidage Profile.
U_g=3.8m/s, G_s=98kg/m²s, Nozzles at the wall.

Figure 4 shows the corresponding radial density profiles at two elevations, 0.80m and 1.50m, above the nozzle inlet. It is seen that the nozzle configuration has an important influence on the radial solid density profile. When the nozzles are at the center of the riser, there is a

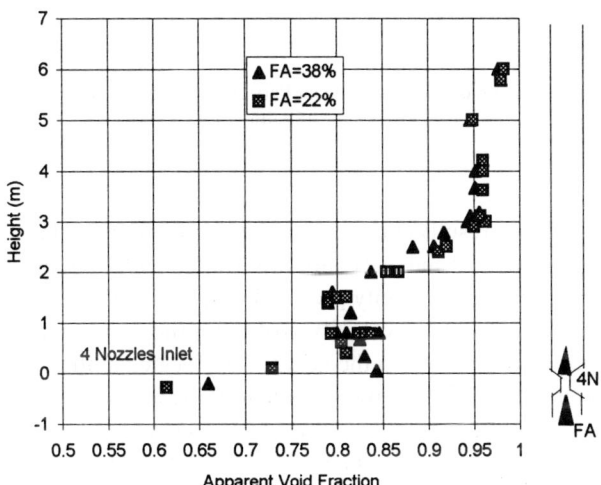

Figure 2: Axial Voidage Profile.
U_g=3.8m/s, G_s=111kg/m²s, Nozzles at the center.

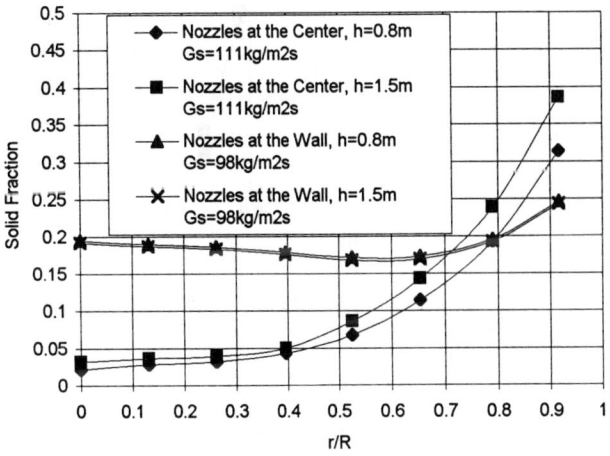

Figure 4: Radial Solid Density Profile.
U_g=3.8m/s, 4N=78%, FA=22%

strong radial concentration gradient with solid fraction very low at the riser axis and very high close the riser wall. When the nozzles are moved close the riser wall the radial profile appears almost flat along the riser radius. The radial profile for the nozzle at the riser center at 1.50m is greater than the profile at 0.80m. Applying solid continuity with this observation shows that the solid decelerates from 0.80m to 1.50m when the nozzles are located at the center of the riser.

Figure 5 shows the radial profile of the axial net solid flux for superficial gas velocity U_g=3.8m/s and solid circulation rate G_s=148kg/m²s. The "fluffing" air flowrate is 22% of the total flowrate. The results show that the solid flux profile does not change significantly with elevation over this height for either of the two nozzle configurations. When the nozzles are located at the center, the flux profile has an almost parabolic shape with negative values close to the wall (from r/R=0.85 to r/R=1) and with a maximum at the riser axis. When the nozzles are located at the wall, the flux profile becomes flatter at the section close the riser axis. The total solid circulation rates obtained from the integration of the net solid flux profile are also presented in Figure 5 for each case and it can be seen that they are in very good agreement with the measured solid circulation rate.

located at the center the solid velocity profile has a strong gradient in the radial direction as does the radial density profile. The solid velocity rises from negative values close the riser wall to about 7m/s at the riser center. The Figure clearly shows the deceleration of solid that takes place from 0.75m to 1.70m. The axial solid velocity decreases over 60% of the riser radius with maximum deceleration at the riser axis (from about 7m/s at 0.75m to about 5.5m/s at 1.70m). This deceleration of the solid combined with the strong radial density gradient indicates poor gas-solid contact and mixing and a high degree of gas backmixing. When the nozzles are located at the riser wall the velocity profile appears flat with lower velocities since the corresponding solid fraction is high. There is no deceleration between the two elevations indicating that this nozzle configuration is much better than the other one.

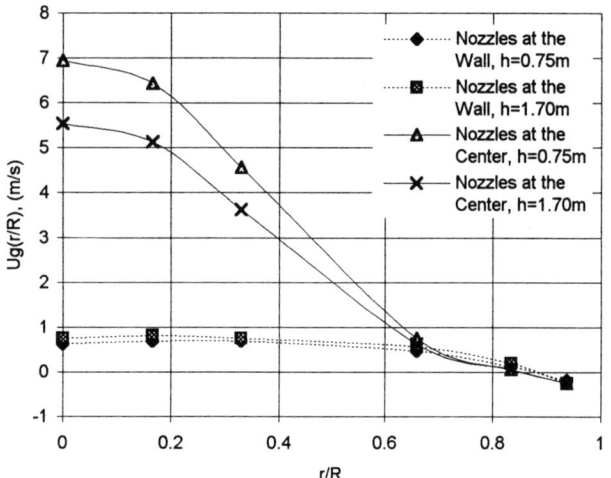

Figure 6: Calculated Axial Particle Velocity Profile. U_g=3.8m/s, G_s=148kg/m²s, FA=22%.

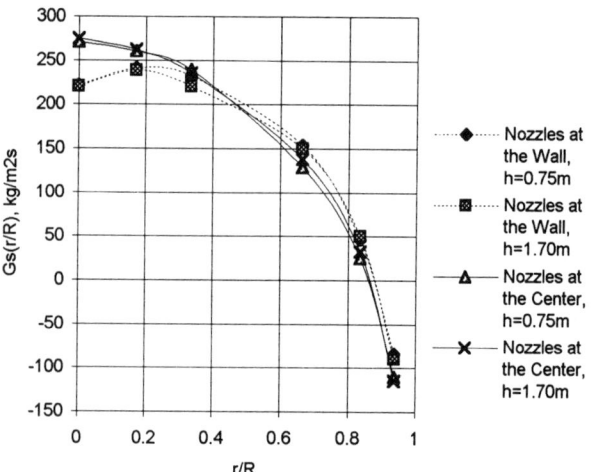

Figure 5: Solid Flux Profile. U_g=3.8m/s, G_s=148kg/m²s, FA=22%, U_{asp}=2.5m/s.

From these results, the radial density profile and the radial profile of the axial net solid flux, a radial profile of the axial solid velocity can be calculated. The results are shown in Figure 6. It is seen that when the nozzles are

These results combined with a correlation for the slip velocity [3] and gas continuity can give an estimation of the axial gas velocity profile. Figure 7 shows the computed axial gas velocity profile for the two nozzle configurations. When the nozzles are located at the center the profile shows a gradient of the axial gas velocity with maximum value at the riser axis. Gas velocity decreases from r/R=0 to r/R=0.45 and increases over the rest of the radius from an elevation of 0.75m to 1.70m. On the other hand when the nozzles are located at the riser wall the gas velocity profile is flat over the riser radius as are the density and solid velocity profiles.

There is not any significant difference in the profile from 0.75m to 1.70m.

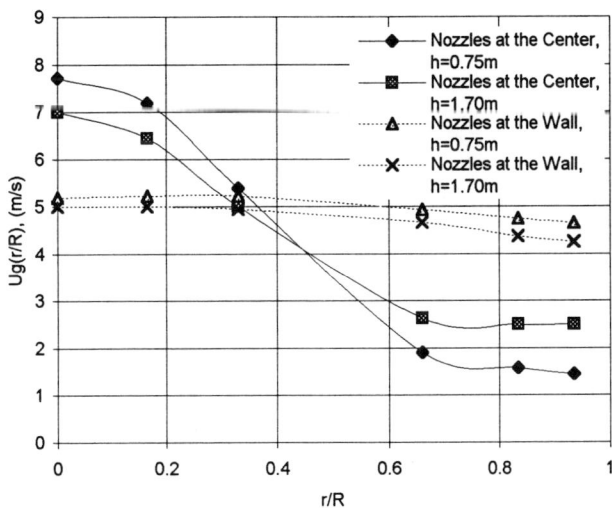

Figure 7: Estimated Axial Gas Velocity Profile.
U_g=3.8m/s, G_s=148kg/m^{2}s, FA=22%.

CONCLUSIONS

◆ Nozzle configuration and "fluffing" air flowrate both significantly affect the gas-solid flow in the riser and therefore its performance as chemical reactor.

◆ Higher "fluffing" air flowrates give smoother axial apparent voidage profiles when the nozzles are located at the center of the riser. "Fluffing" air flowrate does not affect the axial voidage profile when the nozzles are located as close as possible to the riser wall.

◆ The entrance configuration with the nozzles at the center shows strong radial gradients. These strong gradients imply poor performance of the riser as a reactor. On the other hand, the entrance configuration with the nozzles as close as possible to the riser wall give flat radial distributions. These flat profiles indicate very good gas-solid mixing and minimization of the gas backmixing which should result in higher reaction selectivity and yield and therefore higher reactor performance.

ACKNOWLEDGMENT

This work was Supported by the National Science Foundation under Grant Number CTS-9312099.

LITERATURE CITED

1. Tarmy Barry L., "Reaction-Path Synthesis", Reprinted from Kirk-Othmer: ENCYCLOPEDIA OF CHEMICAL TECHNOLOGY, Volume 19, Third Edition, Copyright 1982 by John Wiley & Sons, Inc.

2. Feindt Hans-Jacob, "Radial and Axial Density Fluctuations in a High Velocity Fluidized Bed", Ph.D. Thesis (1990), The City University of New York.

3. Matsen J.M., " Flow of Fluidized Solids and Bubbles in Standpipe and Risers", Powder Technology, 7 (1973), pp. 93-96.

Solids Residence Time Distribution in Interconnected Fluidized Beds

C.S. Stellema[1,2], Z.I. Kolar[1], J.J.M. de Goeij[1], J.C. Schouten[2], and C.M. van den Bleek[2]

[1]Interfaculty Reactor Institute, Delft University of Technology, Mekelweg 15, Delft, The Netherlands
[2]Faculty of Chemical Technology and Materials Science, Delft University of Technology
Julianalaan 136, Delft, The Netherlands

A single-particle radiotracer method has been successfully used to determine the solids residence time distribution (RTD) in a gas-solids Interconnected Fluidized Beds (IFB) setup. The RTD of the risers hardly showed any deviation from the RTD of an ideal mixer. From the RTD of the downcomers important insight in the downcomer solids flow behavior was obtained.

An Interconnected Fluidized Beds (IFB) system consists of a series of risers and downcomers connected to each other by orifices or weirs. Differences in aeration of the beds results in a solids flow from the risers over the weirs into the downcomers and in a solids flow from the downcomers through the orifices back to the risers. Since 1989 IFB setups have continuously been subject of investigation at Delft University of Technology.

Applications of IFB technology are in the continuous gas/solids acceptation/regeneration processes such as in-situ regenerative desulfurization during fluidized bed coal combustion [1] or selective oxidation of n-butane [2]. The advantages of IFB systems as compared to conventional solids circulation systems are reactor compactness and absence of solids break-up. As there is no need for pneumatic conveying, particle velocities do not exceed regular fluidization values ($\sim 1 \mathrm{m \cdot s^{-1}}$).

In multiphase reactors, the flow and contacting of the phases is of major influence on the overall reactor performance. As part of this, the circulation rate of solids (CRS) has earlier been investigated and modelled by Korbee *et al.* [3]. In this paper, the attention is focused on the solids residence time distribution (RTD). Apart from the residence times and associated CRS, the RTD may also provide essential information on the solids flow behavior. Both solids and gas flow behavior will be the subject of detailed future research in our laboratories.

Several other groups have investigated IFB reactors. Only few of these have used solids tracers to determine solids RTD's. Szentmarjay *et al.* [4] determined the overall solids RTD in an equally fluidized two-bed IFB system by pulse injection of a batch of radioactive particles. Pajongvit and Jovanovic [5] used differently coloured particles in similar setups. They determined also only the overall solids RTD.

In the present RTD-study a single radiotracer particle method has been developed, tested and used. From the behavior of the single particle conclusions on the behavior of the total solid phase were drawn. This method enables the determination of solids RTD's in each individual differently operated bed in an experimental four-beds IFB setup.

---

[1] Interfaculty Reactor Institute, Delft University of Technology, Delft, The Netherlands
[2] Faculty of Chemical Technology and Materials Science, Delft University of Technology, The Netherlands

EXPERIMENTAL SETUP

Materials

The four-beds IFB-system made of perspex is shown in Fig. 1. The bottom of each bed measures 100 mm x 100 mm. Air is fed into each bed by massflow controllers through a sintered bronze distributor plate. The orifices are located at a height of 100 mm above the plate. They are 20 mm deep and have diameters ranging 15-30 mm. The height of the weirs is 300 mm.

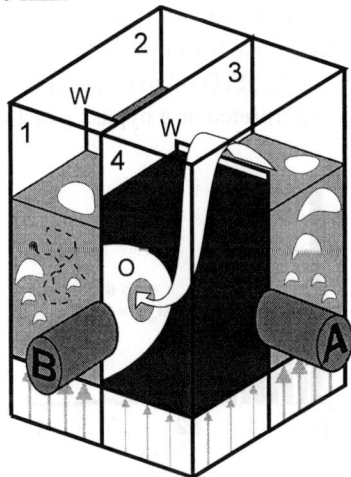

Figure 1. IFB-setup with radiotracer particle (•), solids flow direction and γ-ray detectors.
1,3 : risers; fluidized 2,4 : downcomers; aerated
A,B : γ-ray detectors O,W : orifice, weir

The IFB-setup is filled with a 13-15 kg batch of Geldart B-type glass ballotini (sizes between 550 and 750 μm, see Fig. 3; density 2500 kg·m^{-3}, minimum fluidization velocity 0.36 m·s^{-1}). The inside of the IFB-setup was coated with Statguard® conductive floor finish to eliminate effects of static electricity.

Elemental analysis of the ballotini showed that they contain 0.6% Sb. Several ballotini of different sizes within the particle size distribution were irradiated in the nuclear research reactor with thermal neutrons to obtain so called native radiotracer particles. The time integrated neutron flux amounted to 1.8·10^{24} neutrons·m^{-2}. After 20 days of decay the dominant activity came from ^{124}Sb (around 1.5 MBq per particle, mainly 603 keV γ-rays, half life 60 days).

Two 3"x3" NaI(Tl) scintillation detectors, coupled to single channel analyzers, are mounted on the outside of the IFB-system. One detector (A) is directed at the wall between bed 3 and 4 and the other detector (B) at the orifice between bed 4 and 1. The energy window of the analyzers is set from 550 to 650 keV to detect unscattered radiation mainly. The single channel analyzers are connected to a specially designed PC mounted data acquisition card. It continuously registers the counts collected during successive preset time intervals (dwell time). Depending on the CRS, the dwell time was set between 250 ms and 1 s.

Methods

At the start of each experiment a single radiotracer particle is put into the IFB. The dose equivalent for an experimenter at a distance of 1 m from the setup always stayed below 1 μSv·hr^{-1}.

The curves of detector response versus time are measured for at least one day. In Fig. 2. the characteristic shape of these curves is illustrated.

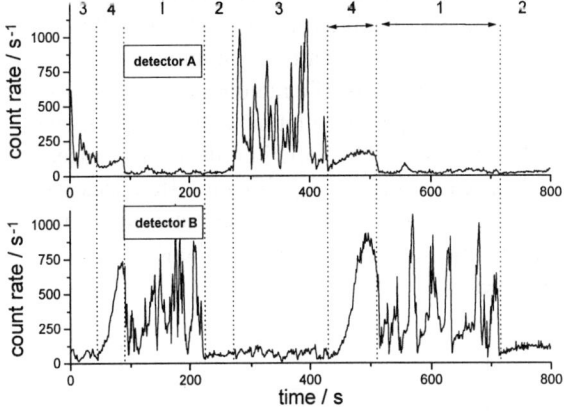

Figure 2. Typical response of detectors A and B. The afterwards determined residence time intervals corresponding to beds 1,2,3 and 4 are also shown.

From Table 1. it can be learned how the response of detector A and B (see Fig. 2.) translates into the corresponding location of the radiotracer particle.

detector response		tracer location	
A	B	bed	bed function
low	high	1	riser
low	low	2	downcomer
high	low	3	riser
high	high	4	downcomer

Table 1. Average detector response when the radiotracer particle is in one of the four beds.

The criterion for the identification of a switch of the radiotracer particle to another bed lies in the large relative change in the response of the detector that is in line with the movement (alternatingly). Strong fluctuations in the signal for bed 1 (detector B) and for bed 3 (detector A) indicate a rather random movement, while a more gradual course in bed 4 indicates a rather smooth movement (detector A and B), particularly when aerated below the minimum fluidization velocity. The response of bed 2 is weak, since bed 2 is not well 'seen' by either detector A or detector B. The resulting residence time of a particle in a particular bed is determined afterwards by software that was specially developed for this purpose.

The thus collected residence times are stored in normalized and dimensionless histograms which represent discrete steady state solids residence time distributions. With this method residence time distributions which have a relative uncertainty in the second moment of approximately 15% (directly computed from the data) are very common.

The representativity of a radiotracer particle with a certain particle size for ballotini with a distribution of sizes has been tested by variation of the radiotracer particle size (see Fig. 3). Significant differences are neither found in the mean residence times nor in the second moment of the RTD's or in the RTD's themselves.

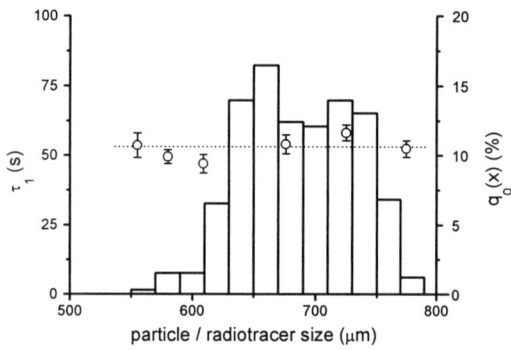

Figure 3. *Mean residence times in bed 1, $\tau_1(s)$, with 66% confidence intervals while the radiotracer particle size varied within the particle size distribution, $q_0(x)$*

The fluidization rates in the risers were set between 1.8 and 2.5 U_{mf}. The downcomers were aerated between 0.7 and 1.3 U_{mf}. Earlier experiments by Korbee et al. [3], in which the solids flow through an open IFB system was measured, showed that in this regime the aeration rate of the downcomer provides an excellent mean to control the CRS.

RESULTS

The solids RTD of the risers, from a chemical point of view the more interesting, is in most cases that of an ideal mixer. The assumption that a riser is an ideal mixer is supported by the fluctuations of the detector signals, indicating a high internal circulation when the radiotracer particle is in a riser (see Fig. 2), *cf.* bed 3, detector A and bed 1, detector B. Only at very high CRS (above 0.15 kg·s^{-1}), significant deviations from the RTD of an ideal mixer occur which can be interpreted as a bypass (see Fig. 4).

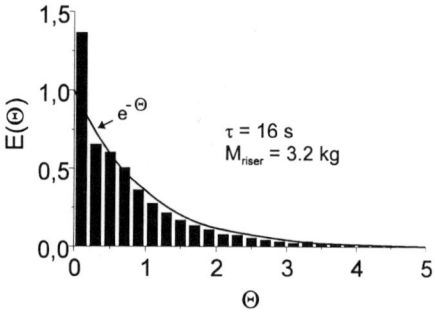

Figure 4. *Riser RTD showing a bypass*

The attention therefore is focused on the residence times in the downcomers. Previous experiments with coloured particles in an externally fed open two-beds IFB system indicated the presence of dead zones in the downcomer. Dead zones were not found in the riser. From the residence times in the downcomer the size of the dead zones can be determined and indications of the flow pattern can be found. This information is important as the conditions in the downcomer are of great influence on the CRS and consequently on the overall reactor performance [1].

In this paper four sets of experiments will be discussed. Starting from two sets of downcomer conditions ($D_{orifice}$=15 mm, M_{down}=3.75 kg, U/U_{mf}=0.96 and $D_{orifice}$=30 mm, M_{down}=3.65 kg, U/U_{mf}=0.96) (a) the aeration rate is increased at a constant bed mass and (b) the bed mass is increased at a constant aeration rate. As the bed mass is a dependent variable in an IFB system a new overall steady state was set for each bed mass by adjusting the fluidization rate of the riser(s) and the total IFB mass.

The residence time sequences (see Fig. 2) were

subjected to Spearman and to Kendall rank correlation tests to see whether any correlation exists between the residence time in a downcomer and the residence time in the preceding riser. In none of the experiments a correlation was found. An important consequence of this is that individual downcomer average residence times as well as RTD's measured under different overall IFB conditions can be compared to each other. Another consequence is that the sum of the variances of the RTD's equals the variance of the overall circulation time.

To determine the size of the dead zone(s) in a downcomer the ratio of the downcomer bed mass (determined after an abrupt stop of the air feed with an accuracy of approximately 2 %) and its mean residence time was compared to the CRS. In steady state the CRS equals the ratio of the bed mass in a riser and its mean residence time.

Fig. 5 and 6 show that increasing the aeration rate from the starting conditions as well as increasing the bed mass will increase the CRS.

From Fig. 6 it can be seen that the solids flow through the 30 mm orifice is approximately a factor 4 higher than through a 15 mm orifice as is illustrated by the slopes of the dotted lines. This implies that the solids flux is roughly independent of the orifice size. A similar observation can not be made in Fig. 5 due to different downcomer bed masses.

Fig. 7 and 8 show that the size of the dead region in the downcomer is almost exclusively a function of the aeration rate. As there is a gas leak from the downcomer to the riser through the orifice [3], the dead region in Fig. 7 does not vanish until there is enough gas fed to the bed to fluidize it completely. Apparently this is the case at U/U_{mf}=1.2. Fig. 8 shows that the dead mass remains constant for increasing bed mass at constant aeration.

Fig. 9 to 12 show all the normalised solids RTD's corresponding to the points in Fig. 5 to 8, originating from the four set of experiments.

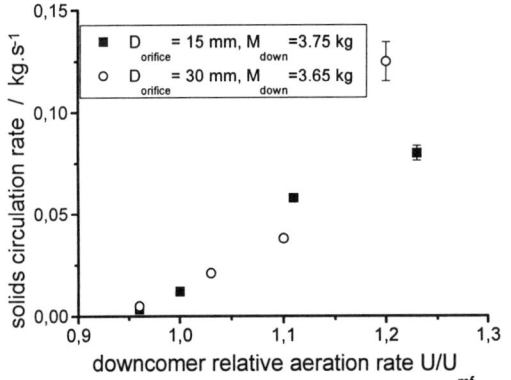

Figure 5. CRS at increasing aeration rate with 95% confidence intervals.

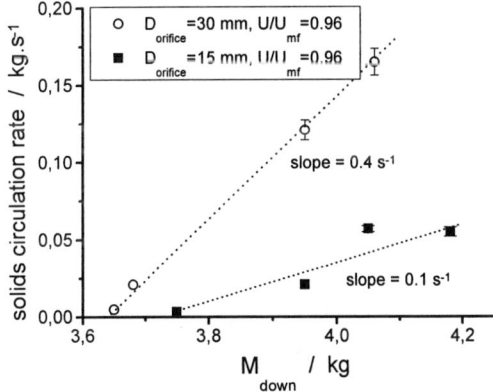

Figure 6. CRS at increasing bed mass with 95% confidence intervals.

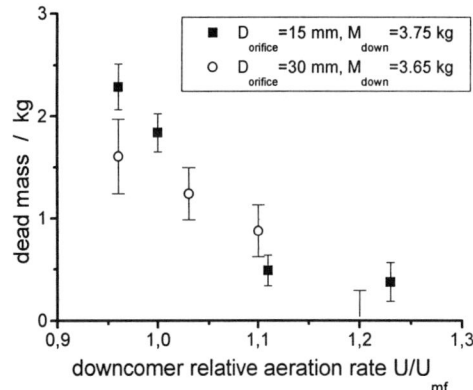

Figure 7. Dead mass in downcomer at constant bed mass with 95% confidence intervals.

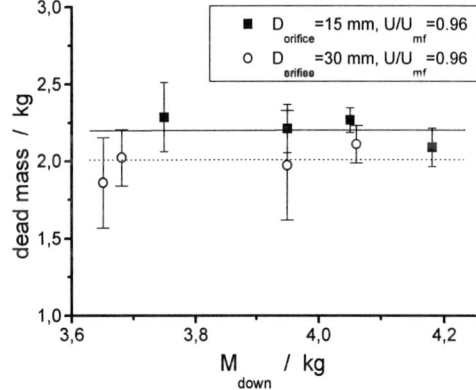

Figure 8. Dead mass in downcomer at increasing bed mass with 95% confidence intervals.

DISCUSSION

The interpretation of Fig. 9 to 12 needs a *caveat*. The first RTD's of each row in these figures are from the two starting experiments. Although the distributions are similar to the RTD of a well mixed tank, visual observation during operation clearly indicates nearly plug flow type behavior. A second indication for a low degree of mixing is the absence of large fluctuations in the detector response curves of Fig. 2 when the radiotracer particle is in bed 4. The small fluctuations present are largely due to the statistical nature of the collected counts.

Similarly, the first and the last RTD in Fig. 9 were not significantly different according to a Kolmogorov Smirnoff two sample test. However, the downcomers in Fig. 9 showed during operation visually different mixing behavior as the last downcomer was fluidized and the first clearly was not. As can be seen in Fig. 7 the active mass increased considerably when U/U_{mf} was increased.

It is likely that the spread in residence times in the first RTD's of Fig. 9 to 12 is due to the different entering positions of the particles in the downcomer after they are ejected from the riser. A different entering position will mean a different trajectory and/or a different velocity. This assumption will be verified experimentally in the near future.

The RTD's in Fig. 11 and 12 show an increasing plug flow character. While the dead region's size in the downcomer remains the same (Fig. 9) the narrowing RTD indicates a decreasing dependence of the residence time on the entering position. For a 30 mm orifice, the narrowing takes place at a lower bed mass.

As indicated above, the described system can locate the single radiotracer particle in one of the four beds as a function of time. However, almost no information as to the position of the particle in a given bed is obtained. Positron emitting radiotracer particles enable to measure 3D distributions, similarly to 3D imaging with PET cameras in nuclear medial diagnosis. In our laboratory a positron emission particle tracking setup has recently become operational. It will enable to follow the 3D path of a positron emitting radiotracer particle. This may lead to an increased understanding of transport and mixing phenomena in downcomers and risers.

NOTATION

τ	average residence time, s
$q_0(x)$	particle size (number density) distribution
Θ	dimensionless residence time
$E(\Theta)$	residence time distribution
M_{riser}	bed mass in riser, kg
M_{down}	bed mass in downcomer, kg
$D_{orifice}$	orifice diameter, mm
U	superficial gas velocity, m·s^{-1}
U_{mf}	minimum fluidization velocity, m·s^{-1}
IFB	Interconnected Fluidized Bed
CRS	Circulation Rate of Solids
RTD	Residence Time Distribution

LITERATURE CITED

1. **Snip, O.C., Korbee, R., Schouten, J.C., and van den Bleek, C.M.,** "The influence of hydrodynamics on the performance of an Interconnected Fluidized Bed System for regenerative desulfurization in coal conversion processes" *AIChE Symposium Series*, **91**, No. 308, 82 (1995).

2. **Maaskant, E.A.R., and Verlaan, M.L,** "The production of THF with the Interconnected Fluidized Bed reactor from butane and air", Internal report, Delft University of Technology (1994).

3. **Korbee, R., Snip, O.C., Schouten, J.C., and van den Bleek, C.M.,** "Rate of Solids and Gas Transfer via an Orifice between Partially and Completely Fluidized Beds," *Chem. Eng. Sci.*, **49**, no. 24 B, 5819-5832 (1994).

4. **Szentmarjay, R., Csukas, B., and Ormoz, Z.,** "Hydrodynamical Studies on Fluidized Beds, part 4: Studies on the Particle Mixing Phenomena in Dual-cell Fluidized Beds using Isotope Tracer Techniques", *Hung. J. Ind. Chem.*, **5**, 213-224 (1977).

5. **Pajongvit, P., and Jovanovic, G.,** "Residence time distribution of solids in a multicompartment fluidized bed system", in: Fluidization and Fluid-Particle Systems Preprints of the American Institute of Chemical Engineers' Annual Meeting, San Francisco (1994).

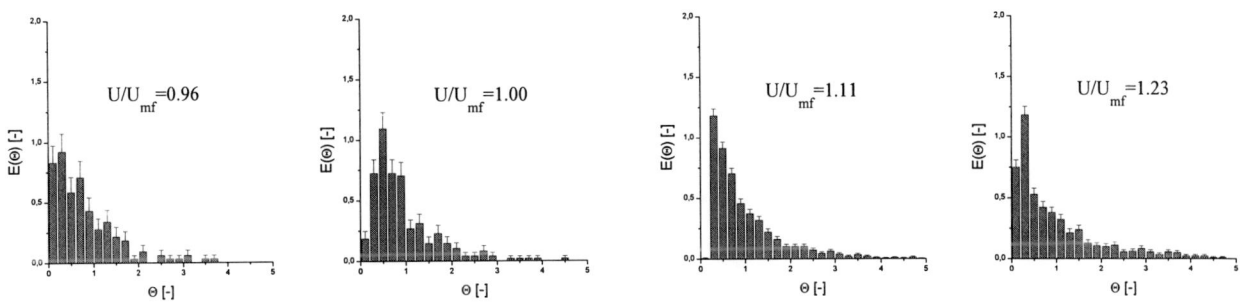

Figure 9. Downcomer RTD's, $D_{orifice}$=15 mm, M_{down}=3.75 kg, increasing U
RTD's are corresponding to Fig. 5 and 7.

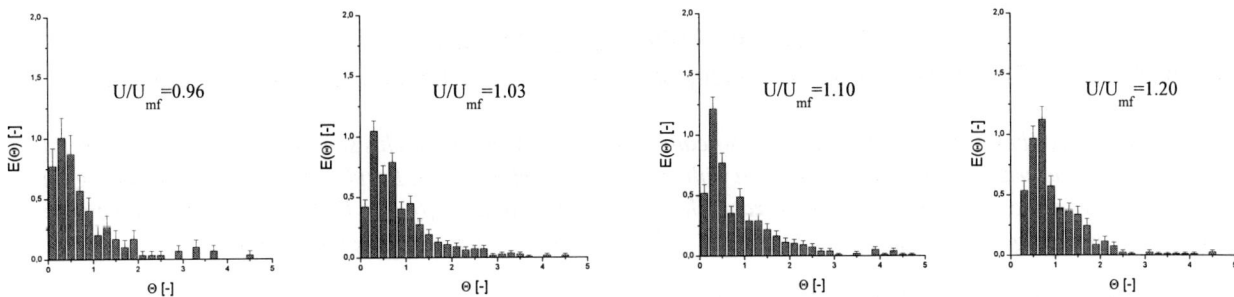

Figure 10. Downcomer RTD's, $D_{orifice}$=30 mm, M_{down}=3.65 kg, increasing U
RTD's correspond to Fig. 5 and 7.

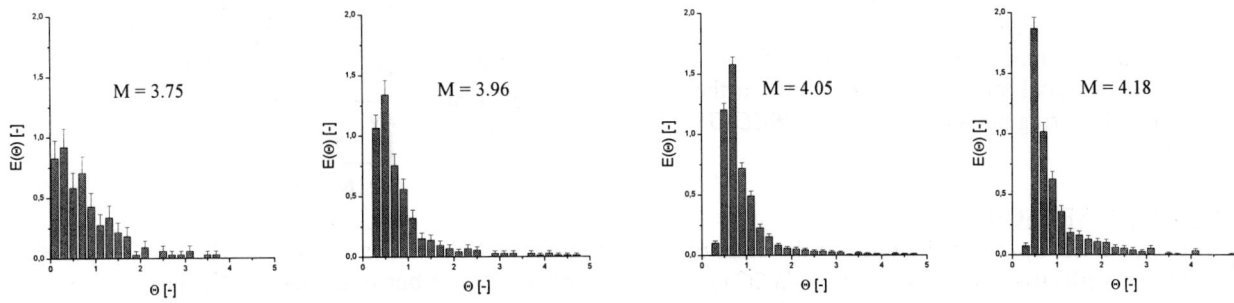

Figure 11. Downcomer RTD's, $D_{orifice}$=15 mm, U/U_{mf}=0.96, increasing M_{down}
RTD's correspond to Fig. 6 and 8.

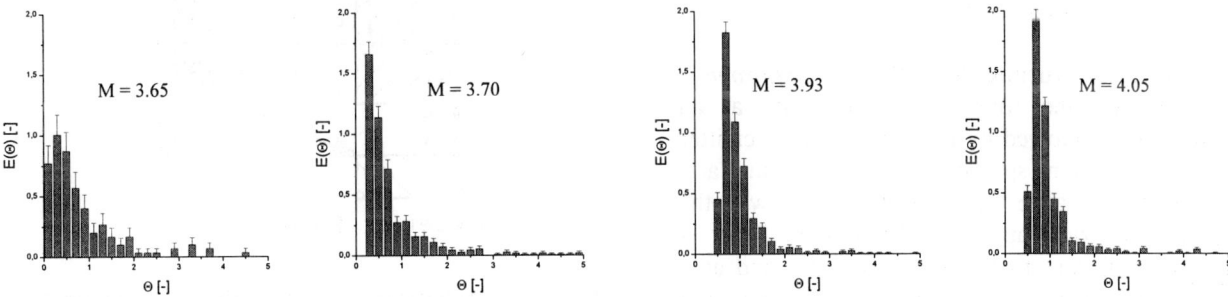

Figure 12. Downcomer RTD's, $D_{orifice}$=30 mm, U/U_{mf}=0.96, increasing M_{down}
RTD's correspond to Fig. 6 and 8.

Particle Clustering in Down Flow Reactors:
Application of a Novel Fiber Optic Sensor

S. Krol, A. Pekediz, H. de Lasa
Chemical Reactor Engineering Centre, University of Western Ontario, London, Ontario N6A 5B9, Canada

This study describes a new optical sensor, the so-called CREC-GS-Optiprobe. *This sensor can be used in gas-solid reactors for characterizing clustering phenomena with minimum intrusion effects. Results show a narrow focal region at 5mm away from the sensor tip. It is demonstrated that in down flow systems (60 μm FCC particles) catalyst evolve as strings with average sizes comprised between 2 to 6* $d_{p,av}$ *with most of these strings having 3.5* $d_{p,av}$

Particle cluster is a phenomenon of importance in many applications in liquid-solid and gas-solid fluidized beds. There have been, in recent years, an interest on particle clustering in gas-solid systems both from theoretical as well as experimental point of views [1]. In downer units particle clustering may be significant and particles may evolve as strings of solids. It is the goal of the present study further describe this clustering phenomenon using *CREC-GS-Optiprobe*.

FIBER OPTIC SENSOR
The *CREC-GS-Optiprobe* belongs to the class of fiber optic reflective sensors. The *CREC-GS-Optiprobe* is based on the following: a) Optimization of optical characteristics for both receiver and emitter, b) Creation of an illuminated cone and a focal point far enough from the sensor tip thus, minimizing intrusion effects.

In its basic structure the *CREC-GS-Optiprobe* is constituted by the following components: a) an *Emitter Fiber* followed by a *GRIN* lens with emitter fiber and *GRIN* lens spaced one form each other at a well selected distance, b) A *Receiver Fiber* with the adequate numerical aperture. The Emitter Fiber is positioned adjacent to the Receiver Fiber and at a well selected distance. Figures 1 and 2 present a sketch of various components of the *CREC-GS-Optiprobe*.

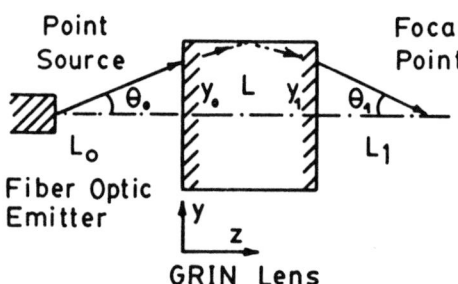

Figure 1. Schematics of the CREC-GS-Optiprobe. Emitter Fiber: fiber optic and GRIN lens.

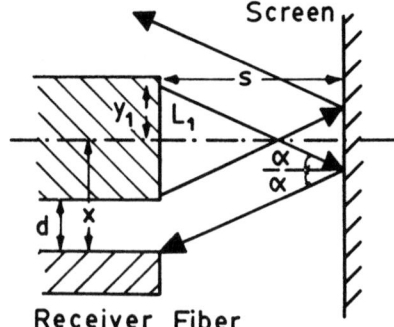

Figure 2. Schematics of the CREC-GS-Optiprobe. Emitter, Receiver Fiber and Plane Screen Simulating Reflecting Particles.

For the first *CREC-GS-Optiprobe* prototype a 3mm tube diameter was chosen to enclose the various sensor components which include the emitter fiber, the GRIN lens and the receiver fiber. The rays emitted from the emitter fiber (point source assumption) reach in the *CREC-GS-Optiprobe* a GRIN lens interface where the Snell law applies (refer to Figure 1).

$$n_{air} \sin[\theta_o] = n_o \sin[\theta_1] \quad (1)$$

Moreover, for rays reaching the GRIN lens interface from air media ($n_{air} = 1$) Equation (1) can be expressed using the paraxial approximation:

$$\theta_o \approx \theta_1 n_o \quad (2)$$

The GRIN lens is a special lens with a graded refraction index [2] changing as follows:

$$n(r) = n_o (1 - A/2 \, r^2) \quad (3)$$

In GRIN lens trajectories of rays can be described with the equation set (refer to Figure 1):

$$y(z) = R_o \sin[\sqrt{A} \, z + \delta] \quad (4)$$

and

$$\theta(z) = \sqrt{A} \, R_o \cos[\sqrt{A} \, z + \delta] \quad (5)$$

R_o and δ depend, in turn, on incident ray parameters θ_o and y_o,

$$R_o = \sqrt{y_o^2 + \frac{\theta_o}{n_o^2 A}} \quad (6)$$

and

$$\delta = \arctan\left[\frac{y_o \, n_o \sqrt{A}}{\theta_o}\right] \quad (7)$$

A GRIN lens of radius R will only accept those rays with $R_o \leq R$. Thus, the extreme rays touching the edge of the lens must have starting values that satisfy the following condition:

$$R = \sqrt{y_o^2 + \frac{\theta_o}{n_o^2 A}} \quad (8)$$

As well, rays evolving in the region before reaching the lens with initial deviation of θ_o direction require a L_o distance to enter the GRIN lens at a y_o height.

$$\theta_o \approx \tan[\theta_o] = [y_o/L_o] \quad (9)$$

Thus,

$$R = \sqrt{y_o^2 \left(1 + \frac{1}{L_o^2 \, n_o^2 \, A}\right)} \quad (10)$$

As a result, the height of the pupil (y_o) can be expressed with the following equation:

$$y_o = \frac{R}{\sqrt{1 + \frac{1}{L_o^2 \, n_o^2 \, A}}} \quad (11)$$

Another important number of considerations concerns the rays in the GRIN lens at the outgoing point (ambient medium side). Let assume that the GRIN lens has the length L and the pitch P. Then,

$$L = 2\pi P/\sqrt{A} \quad (12)$$

and the height of the pupil at the outgoing point (y_1) and the angle of this rays can be expressed as:

$$\theta_1 = \theta_0 \cos(2\pi P) - y_o n_o \sqrt{A} \sin(2\pi P) \quad (13)$$

and

$$y_1 = \frac{\theta_o \sin(2\pi P)}{n_o \sqrt{A}} + y_o \cos(2\pi P) \quad (14)$$

At this point the focal distance (distance where maxmium energy is concentrated), L_1, can be calculated. Given that $\tan(\theta_1)$ can be equated with y_1 and L_1, the focal distance L_1 based on the extreme rays is given by:

$$L_1 = \frac{y_1}{\tan\theta_1} \quad (15)$$

With this information a *Visible Domain* for the receiver fiber can be calculated. In fact, geometrical considerations based on Figure 2 yield the following geometrical parameters,

$$x = (s-L_1)\tan\alpha + s\tan\alpha = (2s-L_1)\tan\alpha \quad (16)$$

with x being the separation between the axis of the emitter and the edge of the receiver core.

As well given that $\tan\alpha = y_1/L_1$:

$$x = (\frac{2s}{L_1} - 1) y_1 \quad (17)$$

and considering the condition of detecting light it results:

$$x \geq y_1 + d^* \quad (18)$$

with $d^* = d + (R - y_1)$

Making the appropriate substitutions the critical distance, s, defining the visible domain is given by:

$$s \geq L_1 (1 + \frac{d}{2y_1}) \quad (19)$$

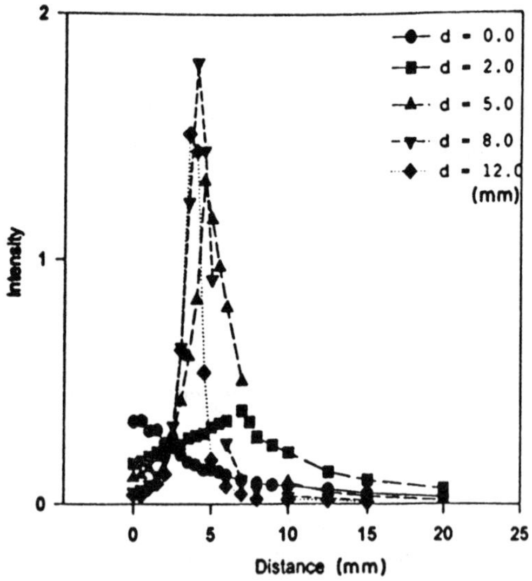

Calculations demonstrate (refer to Table 1) that a most probable focal point calculated for *CREC-GS-Optiprobe* ($L_o = 10$mm) is going to be placed, in the case of extreme rays, at 4.5mm from the GRIN lens surface. This means that all rays coming from a reflecting plane surface placed at 6.1 mm will be "seen" by the Receiver Fiber (visible domain). Other rays, originated from reflecting plane surfaces placed at smaller axial distances, will pass unnoticed. The adequacy of the above calculations was demonstrated in an optical bench and using a sensor prototype. Results reported in Figure 3 show a narrow focal region at 5mm with *CREC-GS-Optiprobe* with the following characteristics: $L_o = 8$ to 12mm for a 0.1 pitch GRIN lens.

CALIBRATION OF THE *CREC-GS-Optiprobe*

To measure clusters/strings particle sizes a calibration *CREC-GS-Optiprobe* was needed. The approach selected uses a rotating disk moving at a well defined angular velocity with strings of 1, 2, 3 and 4 particles held in tracks. This rotating disk was operated with velocities similar to the ones expected in a down flow unit. Figure 4 provides an example of calibrations with the rotating disk set at different positions inside the *CREC-GS-Optiprobe* illuminated volume.

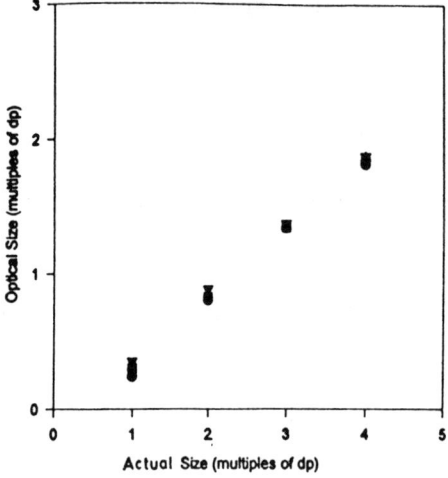

Figure 4. Calibration of CREC-GS-Optiprobe with different disk velocities: (●) 1.5 m/s, (■) 2 m/s; (▲) 2.8 m/s;(▼) 3.3 m/s.

Figure 3. Axial Distribution of light intensity in watts x 10^{-7} for various distances between fiber probe tip and GRIN lens. Pitch of GRIN lens: 0.1. Highest intensity shows focal region.

It was observed from data obtained and cross-correlation analysis that the *CREC-GS Optiprobe* measures so-called "optical sizes" which are about 50% smaller than the actual particle sizes. As shown in Figure 4, the ratio of optical size to actual particle size, was found to be essentialy independent of the disk velocity and of the disk position inside the illuminated volume. Thus, no bias is expected while sizing particle clusters.

PARTICLE CLUSTERING

An intriguing issue for novel downer reactors is the way particles evolve in the unit: single particle or clusters/strings of particles [1].

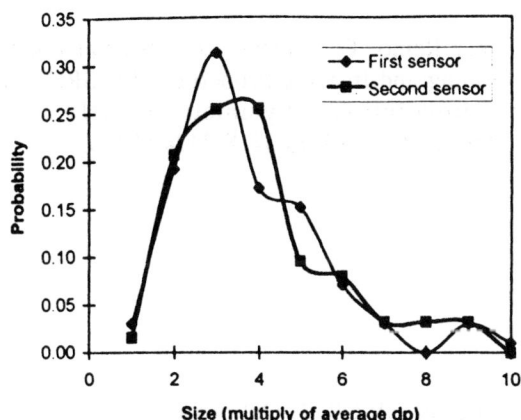

Figure 5. Frequency (probability) of cluster sizes from CREC-GS-Optiprobe in a 2.5 cm downer. Sensor in central position. Spacing between upper and lower sensor spaced 0.6 cm. Solid flux = 8.3 Kg/m².s. Gas superficial velocity: 0.4 m/s.

Figure 5 provides the distribution (frequency) of the ratio of cluster size/average particle size for two sensors spaced axially 0.6cm (lower and the upper sensor) of a downer 2.5 cm diameter unit [1]. Sensors were located at the centerline during this series of measurements. It can be observed that for down flow systems with FCC catalyst (60 μm average), particles evolve as strings with average sizes comprised between 2 to 6 $d_{p,av}$ with most of them having 3.5 $d_{p,av}$. It has to be mentioned that this result was a consistent one for both the upper and lower sensors and at various operating conditions studied. As shown in Figures 6, the same type of particle agglomeration was found in a relatively ample range of solid fluxes 3.5 to 8.5 Kg/m² s with gas superficial velocities set at 0.4 m/s. Moreover, slip velocities comprised between 0.95 to 1.96 m/s were observed (refer to Figure 6). These velocities yield slip velocities several times the terminal velocity of individual particles and this provides additional support to the view that particle clustering takes place in downer reactors with individual particles moving as string shaped agglomerates.

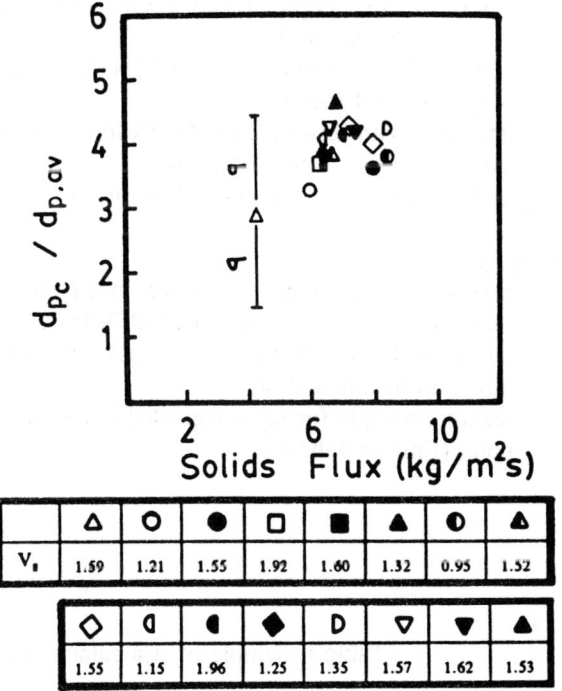

Figure 6. Ratio of the average cluster size over the average particle size with different solid fluxes in a 2.5 downer reactor. Gas superficial velocity = 0.4 m/s.

NOTATION

A GRIN lens gradient constant (1/mm²)
d spacing between the fiber optic light guides (mm)
d^* distance defined as d + (R- y_1))
d_{pc} cluster particle size (cm)
$d_{p,av}$ average size of particles (cm)
L length of the GRIN lens (mm)
L_o distance between the end of the Emitter Fiber and the GRIN lens (mm)
L_1 distance between the end of the GRIN lens and the Focal Point (Image Point) (mm)
n_o refraction index of GRIN lens at r=0
n(r) refraction index at the radial position r
P pitch of the GRIN lens (radians)
r radial coordinate in the GRIN lens (mm)
R radius of the GRIN lens (mm)
R_o maximum rays deviation from centerline (mm)
s distance between the tip of the *CREC-GS-Optiprobe* and the screen (mm)
u_g superficial gas velocity (m s⁻¹)
v_s slip velocity = cluster velocity - superficial gas velocity (m s⁻¹)

x   separation between the axis of the emitter and the edge of the receiver(mm)

y_o height of the pupil at the GRIN lens ray incoming point (mm)

y_1 height of the pupil at the GRIN lens ray outgoing point (mm)

y(z)   deviation with respect to the GRIN lens central axis of a ray at z position (mm)

z   axial coordinate in the GRIN lens (mm)

Greek Symbols

α maximum angle of ray incidence on screen or maximum angle of ray leaving GRIN lens.

δ phase shift determined by a position and direction of incident ray

Θ_o maximum angle for rays originated in a source point (Emitter Fiber).

Θ_1 maximum angle for the rays originated in the GRIN lens at the height y_1 of the pupil

$\Theta(z)$ angle (slope) of a ray evolving in the GRIN lens at z position.

LITERATURE CITED

1. Sobocinski D.J., et al., "New Fiber Optic Method for Measuring Velocities of Strands and Solid Hold-up in Gas-Solid Down Flow Reactors", *Powder Technology*, **83**, pp 1-11 (1995).

2. Gómez-Reino, C., Liñares, J.,"Paraxial Fourier Trasforming and Imaging Properties of a GRIN lens with Revolution Symmetry:GRIN Lens Law" *Appl.Opt.*, **25**(19) pp 3418-3424 (1986).

Table I. Sample of Calculations and Characteristics of GS-CREC-Optiprobe

GRIN lens Parameters	$n_o = 1.6075$; $A = 0.092416$ (1/mm^2); $P = 0.1$; $R = 1.0$ mm
Geometry of *CREC-GS-Optiprobe*	$d = 0.65$ mm
Additional Geometrical Parameters	$L_o = 10$ mm
Rays at GRIN lens-incoming pupil	$y_o = 0.98$ mm; $\theta_o = 4.51$ degrees
Rays at GRIN lens-outgoing pupil	$y_1 = 0.91$ mm; $\theta_1 = 11.57$ degrees
Focal Point	$L_1 = 4.5$ mm
Visible Domain	For $d^* = 0.75$ mm; $s = 6.1$ mm.

Gas-Liquid Mass Transfer in Three-Phase Inverse Fluidized Bed

Ihab. H. Farag and Volen. R. Nikolov
Department of Chemical Engineering, University of New Hampshire, Durham, NH 03824

Iordan Nikov
Institute of Chemical Engineering, Bulgarian Academy of Sciences, Sofia 1113, Bulgaria

In the present study the oxygen mass transfer from the gas to the aqueous phase in a Three-Phase Inverse Fluidized Bed (TPIFB) has been studied. A pilot scale TPIFB has been designed and constructed. For determination of the volumetric oxygen mass transfer coefficient, the elegant dynamic method, described by Dang et al., (1977) was used. The influence of hydro-dynamic parameters, e.g., superficial velocities of the gas and liquid phases on the mass transfer rate was studied. In the range of variables covered, it was found that the superficial liquid velocity had a weak effect on the mass transfer whereas the gas flowrate affects the mass transfer positively. The results for the volumetric oxygen transfer coefficient in the TPIFB were compared to reported values of that coefficient, measured in a classic Three-Phase Fluidized Bed under similar hydrodynamic conditions and solid phase properties. The comparison demonstrated a two-fold increase of the oxygen transfer rate in the inverse bed over that in the classic one.

Fluidized Bed systems have proved their versatility for carrying out aerobic fermentation processes and bio-treatment of wastewater. In designing these bio-treatment operations a key parameter is the volumetric oxygen mass transfer coefficient, denoted by $K_L a$. Traditionally, fluidized beds used in practice consist of particles denser than the liquid phase. The particle bed is expanded upward by cocurrent up-flows of gas and liquid, both acting against net gravitational force of the particles. A relatively new and insufficiently studied yet, and hence not abundant in the practice is the Three-Phase Inverse Fluidized Bed (TPIFB). It consists of particles lighter than the liquid phase. The bed expands downward, dragged by down-flowing liquid phase opposite to the net buoyancy force of the particles. The third phase is gas which is introduced as bubbles at the bottom of the TPIFB and rises countercurrently to the liquid flow. A similar kind of unit applied for wastewater treatment has been patented in 1981 by Shimodaira et al. [1]. Works which discuss for the first time the hydrodynamic properties of the TPIFB are [2, 3, 4]. No data on the gas-liquid mass transfer properties of the TPIFB have been found in the literature to date. Because of the drag action of the liquid phase on the gas bubbles it was expected that the gas phase holdup and mean residence time in TPIFB should be significantly higher than in any other reactor configuration known so far. Indeed Legile et al. [4] reported that the gas fraction in the fluid phase in the TPIFB is much larger than that in bubble columns, and at certain superficial velocities of the gas loading, that fraction reaches 50%. Then the gas phase becomes the continuous phase in the fluidized bed. At the same hydrodynamic conditions in a bubble column the authors give a value for the gas fraction of about 15%. This feature should ultimately increase the gas-liquid mass transfer rate and decrease the aeration expenses for aerobic wastewater treatment operations carried out in the TPIFB.

The present work is a study of the oxygen mass transfer in the Three-Phase Inverse Fluidized Bed.

EXPERIMENTAL

The experimental setup is presented in Figure 1. The experiments have been carried out in a acrylic column (1) with a diameter 0.102 m (4 in) and a height 2 m. The column was equipped with a shower head (3) to distribute the liquid phase, with six pressure taps (13) equally spaced every 0.305 m (1 ft) and with a dissolved oxygen (DO) probe *Ingold* (4). A screen (9) was placed beneath the gas distributor (7) to prevent the packing material from penetrating in the circulation centrifugal pump (11) upon drainage of the unit. The packing material was thermally expanded polystyrene beads (2) -- raw material for *Styrofoam* production. This kind of material was used for the first time by Karamanev and Nikolov [5] in their study of the Liquid-Solid Inverse Fluidization. The beads were expanded

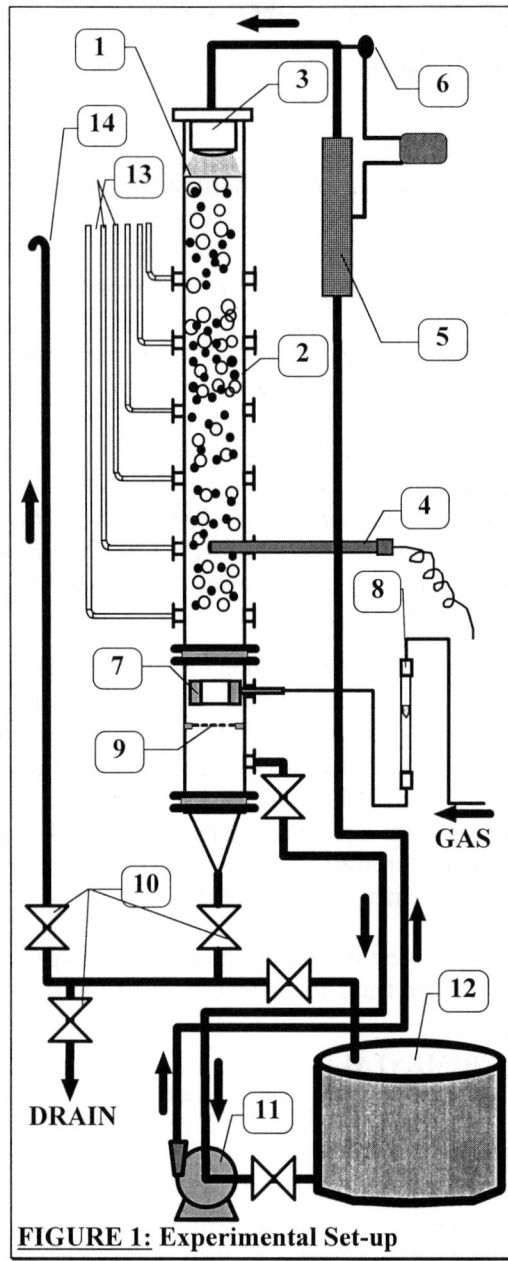

FIGURE 1: Experimental Set-up

with diameters from 1 to 3 mm. It consisted of an annular *Plexiglas* hollow body, capped with *Teflon* foil with 36 perforations of diameter 0.2 mm spaced every 10 degrees on the central circle of the annulus. Such a design did not obstruct the liquid flow and at the same time provided even radial bubble distribution. Teflon was used because of its low contact angle with water, hence easier bubble detachment from the sparger orifices, low coalescence and small bubbles. Filtered air from the compressed air mains, or nitrogen from a 99.99% purity grade gas cylinder were delivered to the gas distributor through a *Gilmont* flowmeter (8). The circulation lines were equipped with globe valves (10), facilitating the experiments and the column had an overflow outlet (14) used for adjustment of the level of the three-phase fluidized zone. All trials to confine the fluidized bed from the top by a supporting grid, as it was done in the works of Fan et al. [2], Fan et al. [3], and Legile et al. [4], were unsuccessful. Even when large mesh screens were used, gas plugs were forming under the grid in result of the surface tension forces. Ultimately this resulted in occasional eruption of the gas plugs through the grid and unstable fluidization.

For determination of the volumetric oxygen transfer coefficient the elegant method proposed by Dang et al. [6] was accordingly modified and used. Initially the first time moment of the probe lag was measured by the rapid placing of the probe in a well mixed aerated water medium after equilibration in nitrogenated gas medium. The electrode response was recorded in intervals of one half of a second. Then the oxygen probe was mounted in the TPIFB. The gas inlet was supplied with nitrogen and the bed was kept at certain liquid and gas superficial velocities until steady-state fluidization and almost zero DO concentration were achieved. Then, instantly, the gas inlet was switched from nitrogen feed to air feed and recording of the electrode response was initiated in intervals of 4 seconds. The flowrate of the air feed was pre-adjusted at the same level as the flowrate of the preceding nitrogen sparging. Thus it was ensured that constant bed expansion and hydrodynamic conditions would be maintained throughout the whole experiment. A program written in BASIC was created to aid the data acquisition and to ensure the proper communication between an IBM PC and a *Solartron* microprocessor multimeter. The gas holdup was calculated on the basis of the pressure drop through the TPIFB. The volume of the solid phase and liquid phase in the

in boiling water to a mean diameter of 3.5 mm and density of 650 kg/m^3. The column was charged (30% of the working volume) with the packing material and then filled with distilled water from the holding tank (12). The heat dissipated from the circulation pump was being removed by the cooling water coil (5) and the temperature was maintained at 30±0.2°C by the temperature controller (6).

A special gas distributor (7) was designed and manufactured, ensuring formation of gas bubbles

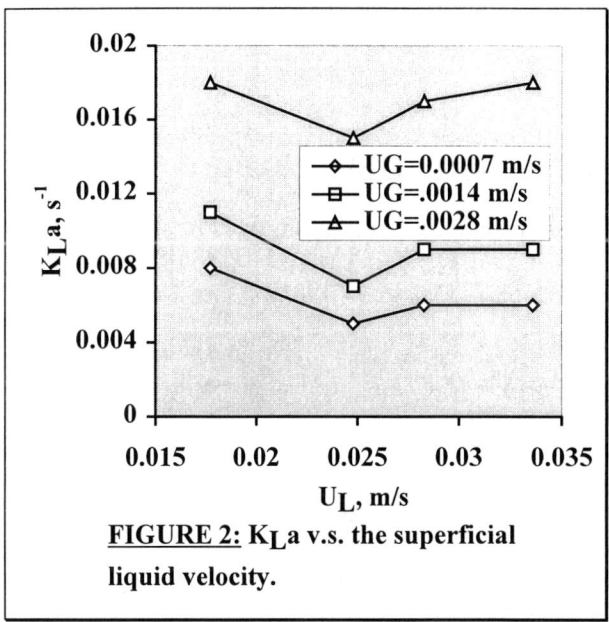

FIGURE 2: K_La v.s. the superficial liquid velocity.

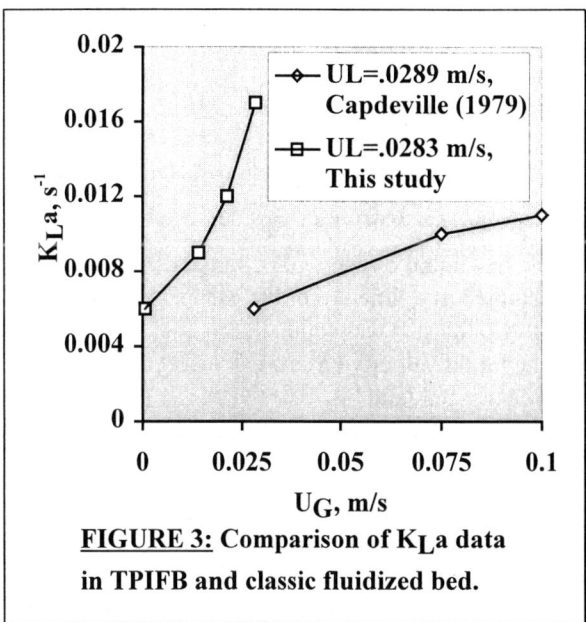

FIGURE 3: Comparison of K_La data in TPIFB and classic fluidized bed.

reactor were measured upon completion of the experiment.

METHOD AND RESULTS DISCUSSION

The volumetric gas-liquid mass transfer coefficients in classic fluidized beds have been measured, in majority, by fitting the axial DO concentration profile to the Axial Dispersion Model [7], to the Plug Flow Model or to the Backflow Cell Model [8]. A common assumption enabling the utilization of these models is: constant oxygen mole fraction in the gas phase. In the TPIFB an even bubbling fluidization could be achieved at superficial gas phase velocities, much lower than those in classic fluidized beds. The lower superficial gas loading and the larger gas holdup in the TPIFB do result in considerable mixing in the gas phase. Thus, it is more reasonable to assume ideal mixing of the gas phase instead. Therefore the methods used for determination of the volumetric oxygen transfer coefficient used for classic three-phase fluidized beds, did not appear to be applicable for the TPIFB. The method used in this study was proposed by Dang et al. [6] for studying oxygen transfer in continuous stirred tank reactors (CSTRs), which are characterized by a considerable mixing in the gas phase.

First in the experimental study, the combined time constant of the oxygen transfer through the electrode membrane and through the liquid film surrounding it, $\tau_F+\tau_E$, was measured under different stirring speeds (Dang et al. [6]). The results showed that at temperature 30°C, solely the electrode membrane constant, τ_E, has a value of about 19 s. The physical meaning of this time constant is the time in seconds, necessary for the electrode response to reach 63.2% of its steady state value, when the DO probe is instantly replaced from one highly turbulent medium to another, which have different DO concentrations. Under vigorous mixing the impedance and capacity of the liquid film, surrounding the electrode membrane could be neglected. This impedance, of course increases as the hydrodynamic conditions become more mild. This resistance is represented by the value of the liquid film time constant, τ_F. Since both resistances are in series, the cumulative time constant should be the sum of the two time constants.

A comparison has been made of the results to similar K_La results obtained in classic fluidization. For the basis of the comparison it was interesting to select experimental data obtained under similar conditions: gas and liquid superficial velocities, density difference between the liquid and solid phases ($\Delta\rho=300$ kg/m^3), and particle diameters ($d_p=3.5$ mm). Such conditions have been found only in the Dissertation work of Capdeville [9]. The comparison (presented in Figure 3) shows that K_La in the TPIFB increases very sharply with the superficial gas velocity and reaches values which are roughly two times higher than the values reported for classic fluidization under similar conditions.

CONCLUSIONS

1. A pilot scale TPIFB was designed and constructed.

2. A special gas distributor was designed.

3. The gas-liquid oxygen mass transfer coefficient was studied as a function of the gas and liquid superficial velocities. It was found that:
- the liquid velocity has a weak effect on the oxygen transfer, and
- the gas velocity affects the oxygen transfer positively.

4. The results were compared to experimental results obtained under similar conditions in classic fluidized beds. It was shown that the oxygen mass transfer in the TPIFB can reach values two times higher than in classic fluidized bed, at lower gas velocities.

ACKNOWLEDGMENT

This study was supported by National Science Foundation as a joint project between Dr. Ihab Farag, University of New Hampshire and Dr. Iordan Nikov, Institute of Chemical Engineering, Bulgarian Academy of Sciences

NOTATION

$K_L a$ - volumetric coefficient of oxygen mass transfer, s^{-1}
U_L - superficial velocity of the liquid phase, m/s
U_G - superficial velocity of the gas phase, m/s
$\Delta \rho$ - difference in the densities of the liquid and solid phases, kg/m^3
τ_E - time constant of the electrode membrane, s
τ_F - time constant of the liquid film surrounding the membrane, s

LITERATURE CITED

1. Shimodaira, C., Y., Yushina, H.H., Kamata, H., Komatsu, and O.M., Jurina, USA Patent 4,256.573,(March 17 1981)

2. Fan, Liang-S., K., Muroyama, and Song-H., Chern, *Chem. Eng. J. and Biochem. Eng. J.,* **24,** 143, (1982)

3. Fan, Liang-S., K., Muroyama, and Song-H., Chern, *Chem. Eng. Sci.,* **37,** 1572, (1982)

4. Legile, P., G., Menard, C., Laurent, D., Thomas, and A., Bernis, *International Chem. Eng.,* **32,** 41, (1992)

5. Karamanev, D.G. and L.N., Nikolov, *AIChE Journal,* **38,** 1916, (1992)

6. Dang, N.D.P., D.A., Karrer, and I.J., Dunn, *Biotech. Bioeng.,* **19,** 853, (1977)

7. Kim, J.O. and S.D., Kim, *Can. J. Chem. Eng.,* **68,** 368, (1990)

8. Nguyen-Tien, K., A.N., Patwari, A., Schumpe, and W.-D., Deckwer, *AIChE Journal,* **31,** 194, (1985)

9. Capdeville, B., Ph.D. Thesis, I.G.C., Toulouse, France (1979)

Capacitance Probe Measurements of Solids Volume Concentrations and Velocities Inside an Industrial Circulating Fluidized Bed Combustor

Bernd Hage and Joachim Werther*
Technical University Hamburg-Harburg, Hamburg, Germany

Although fluidized bed reactors at elevated temperatures (up to 1000°C) have a wide field of application, there is still a lack of measurement techniques for local solids volume concentration and velocity. In this paper, a measurement technique which relies on the measurement of the local dielectricity is described. Through the application of a guard circuit and specially designed probes, an extremely high resolution and stability has been achieved. The calibration function was proved to be linear by comparison of capacitance probe signals to pressure drop measurements in a semi-technical scale circulating fluidized bed (CFB) reactor. The measurement system has been successfully operated under high-temperature conditions in an industrial-scale CFB combustor where the local time-average solids volume concentration was 0.1%.

Up to now, the following measuring principles for flow structure measurements have been tested in hot fluidized bed processes: X-ray absorption, optical reflection and detection of electrical capacitance. However, none of these techniques has become a standard equipment for industrial reactors. Capacitance measurement systems measure the local dielectricity constant of the fluidized bed which has long been known to be linked to the local solids volume concentration. Almstedt and Olsson [1] measured with a needle type two-channel capacitance probe introduced for cold applications by Molerus and Werther [2]. They measured bubble rise velocities in a bubbling fluidized bed combustor by cross-correlation after a suitable drift correction procedure of the signals. The drift encountered in the capacitance signals can easily be explained by the probe setup. The capacitance of the probe body is relatively big in relation to the possible change in capacity caused by particles in the measurement volume at the probe tip. The capacity of the probe body is subject to changes induced by slightly varying temperatures. Since the measured capacity is the sum of the capacitances of the probe body and probe tip, the resulting signal suffers from a temperature drift which can be as big as the measured signal itself.

Technical University Hamburg-Harburg, Hamburg, Germany. B. Hage is now with BP Chemicals SNC, Lavera, France, * to whom correspondence should be adressed

EXPERIMENTAL

Probe Setup

The developed sensor (Figure 1) is a combination of the un-guarded high-temperature probe by Hage et al. [3] and the guarded probe by Soong et al. [4] which was used under ambient conditions. A water-cooled design was chosen as a first development step. The guard electrode is simply an additional coaxial steel tube between the needle electrode and the surrounding ground electrode. The combination of steel and ceramic tubes was chosen with a sufficient spacing for thermal extension. All steel construction elements are made of highly temperature-resistant steel (DIN 1.4841).

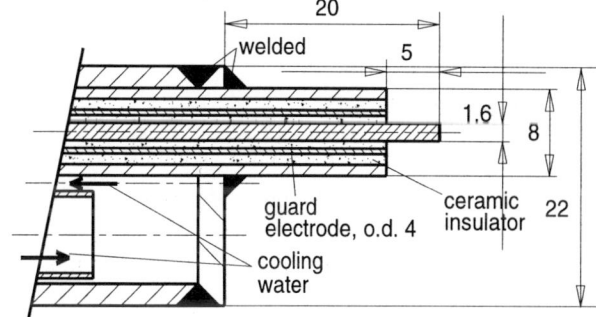

Figure 1. Water-cooled single-channel guarded capacitance sensor for high temperature application (dimensions in millimeters).

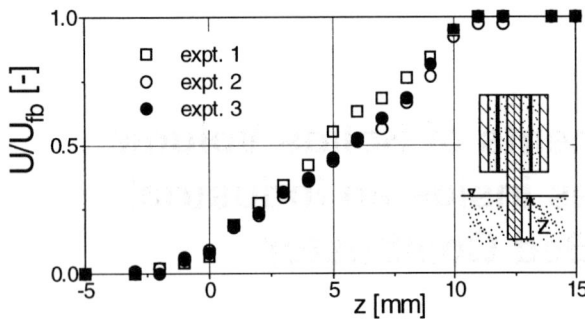

Figure 2. Variation of the dimensionless signal amplitude U/U_{fb} with the depth of immersion x in a fixed bed of sand particles (needle length 10 mm).

Steel connections were carried out by welding. The ceramic tubes were glued in the water-cooled parts of the probe. The simple and relatively big probe setup was chosen in order to provide sufficient mechanical strength against the rough operating and handling conditions in an industrial environment. Probes with total lengths between 990 and 1990 mm were manufactured. The distance of the probe tip from the water-cooled section was chosen to be 15 mm. This length keeps the probe tip at a sufficiently high temperature to avoid condensation of combustion products on its surface. The extension of the measurement volume can be estimated by the signal variation when inserting the probe into a fixed bed (Fig. 2). Horizontal profiles of the local time-averaged vertical solids velocities are obtained by cross-correlation of two signals from a double probe. The probe used for calibration purposes in the semi-technical scale CFB facility had a similar design the main difference being that the outer probe tube had a diameter of only 14 mm taking account of the small reactor diameter of only 100 mm. For temperature measurements in the technical scale reactor, a shielded thermocouple probe was built.

The Fluidized Bed Reactors

The semi-technical scale CFB combustor at the Technical University Hamburg-Harburg (TUHH) provided the possibility of a detailed test of the capacitance probe. The unit a detailed description of which may be found in [5] comprises a riser with a height of 15 m and an inner diameter of 100 mm. The industrial CFB combustor, where measurements were performed with our capacitance probes, is a 109 MW$_{th}$ coal firing boiler (Figure 3). It was erected in 1988 by Lurgi in Flensburg in the north of Germany [6]. This

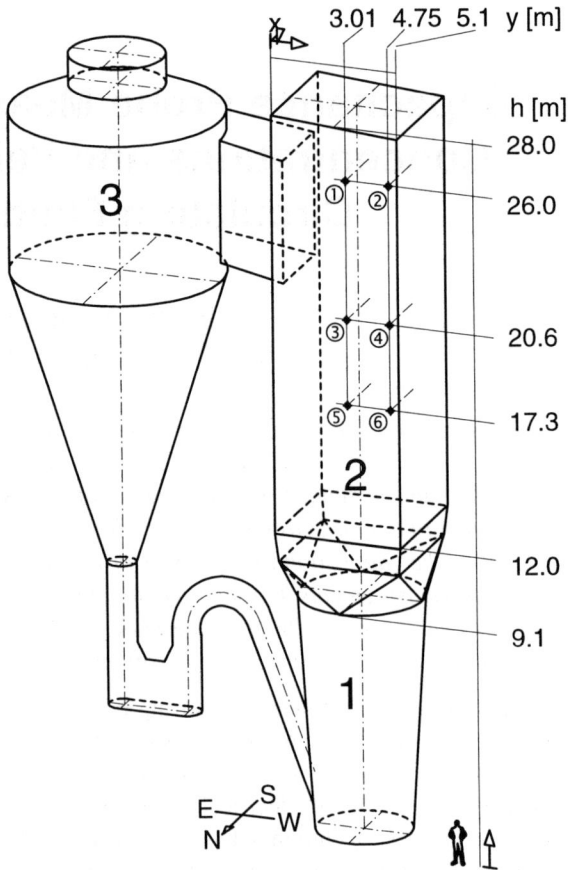

Figure 3 Dimetrical plot of the industrial-scale CFB combustor of the Stadtwerke Flensburg **1**: refractory lined bottom section of the riser, **2**: upper riser section with membrane walls, **3**: cyclone, ①-⑥: measurement ports.

combustor is working as a co-generation plant. The riser has a conical, refractory lined bottom section (3.75 m dia. in the lower part widening to 4.75 m in the upper part of the riser) and a square upper section with a side length of 5.1 m. Figure 3 shows the positions of the measuring ports and the coordinate system. The height is counted from the distributor level, the coordinate x from the north wall of the combustor chamber, and the coordinate y from the east wall. The measurements were carried out at almost 100% load. In the square section of the riser, a superficial gas velocity of 5.9 m/s is calculated when assuming a constant temperature of 800°C over the cross-section of the riser. Previous measurements of local solids mass fluxes with suction probes and temperature profiles were carried out by Werdermann [7] in the same unit.

RESULTS AND DISCUSSION

Calibration of the Capacitance Measurement System

A necessary prerequisite for the calibration of the sensor signals is a low signal drift. A time series of the signal acquired by the guarded measurement system in the wall zone of the Flensburg combustor is shown in Fig. 4.

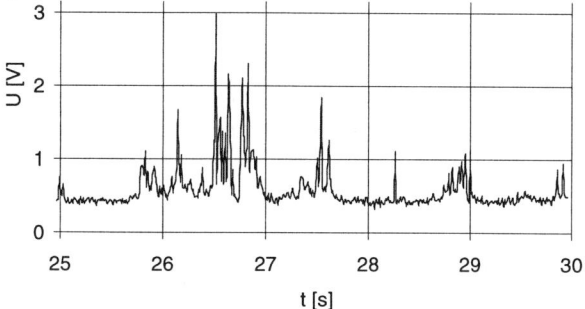

Fig. 4. Guarded capacitance probe signals obtained in the Flensburg combustor (850°C, h=17.3 m, port no.5, x=0, fixed bed signal signal level at 9.4 V).

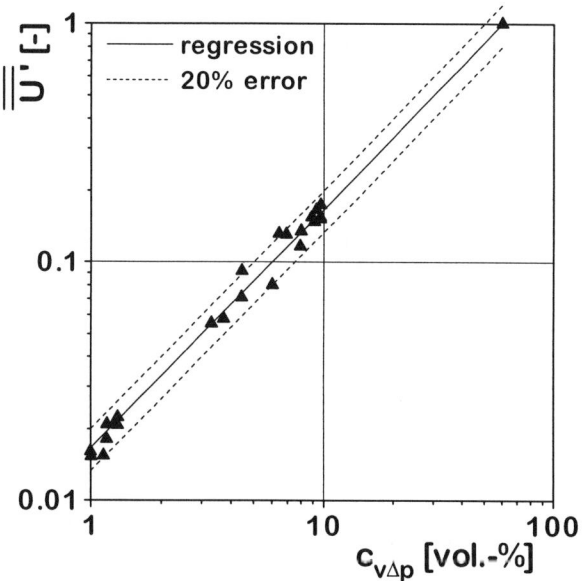

Figure 5. Relation between the average dimensionless signal voltage $\overline{\overline{U'}}$ and the cross-sectional average solids volume concentration obtained from pressure drop (TUHH CFB combustor during sewage sludge combustion, 850°C, h=1.6 m, u=3 -5.4 m/s).

On the basis of the high signal stability achieved, a calibration of the signals became possible. As a simple approach, a linear relationship between the solids volume concentration c_v and the signal voltage U is assumed,

$$c_V(U) = K \cdot (U - U_0). \quad (1)$$

U_0 is the voltage, which corresponds to zero solids volume concentration. Since Equation (1) is assumed to be valid over the full range of possible solid volume concentrations, it also holds for the fixed bed,

$$c_{v\,fb} = K \cdot (U_{fb} - U_0). \quad (2)$$

It follows, $\quad c_v = c_{v\,fb} \cdot \dfrac{U - U_0}{U_{fb} - U_0} = c_{v\,fb} \cdot U' \quad (3)$

Equation (3) defines a relationship between the solids volume concentration and a dimensionless voltage U'. The validity of Equation (3) was examined in the TUHH CFB combustor during sewage sludge combustion. The unit was operated at different operating conditions leading to different cross-sectional solids volume concentrations $c_{v\,\Delta p}$ at the level 1.6 m above the distributor, where the capacitance probe was located. For each operating condition, a traverse was measured with the capacitance probe. The instantaneous dimensionless voltage U' was first time-averaged and then the time averages were averaged again over the cross-sectional area of the riser thus leading to a time and cross-sectional average of the dimensionless signal voltage,

$$\overline{\overline{U'}} = \frac{1}{A_t \cdot n \cdot \Delta t} \sum_{i=1}^{m}(\sum_{j=1}^{n} U' \cdot \Delta t) \cdot A_i, \quad (4)$$

In Figure 5, the dimensionless average signal voltage is plotted against $c_{v\,\Delta p}$. It is clearly seen that a linear relationship exists between this voltage and $c_{v\,\Delta p}$ over a wide range of average solids volume concentrations from 1 to 60 vol.-%, which may be taken as a justification of the linear calibration relationship (Equation (1)).

Measurements in the TUHH circulating fluidized bed combustor

Fig. 6 illustrates the high resolution of the measurement system. Time-averaged c_v as low as 0.1 vol.-% are easily detectable. Fig. 6 shows five cross-sectional measurements traverses, taken under identical operating conditions. Each measurement point corresponds to a time series of one minute, acquired at

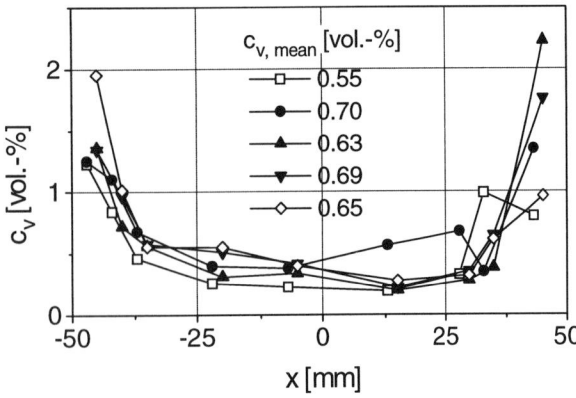

Fig. 6. Solids volume concentration in the upper dilute region of a circulating fluidized bed combustor (TUHH unit, h=11.5 m, $c_{v\Delta p}$= 0.5-0.7 vol.-%, uncooled guarded capacitance probe, all points under identical operating conditions, each data point acquired over 1 min at 100 Hz).

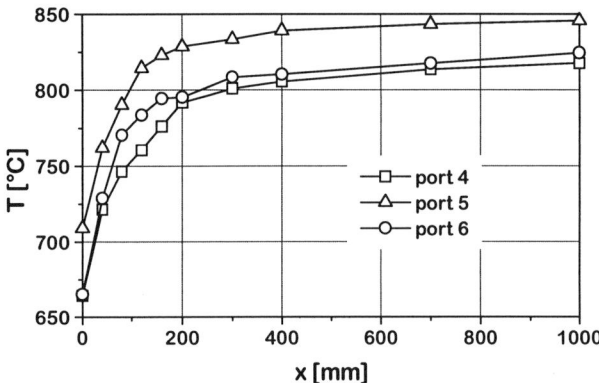

Fig. 8. Temperatures near the combustor wall in the Flensburg combustor (850°C, for port locations see Fig. 3, x is distance from the wall, counted from the fin).

100 Hz. The measurements indicate a good reproducibility of the results. Simultaneously recorded pressure drop measurements yielded cross-sectional solids volume concentrations $c_{v\,\Delta p}$ varying between 0.5 and 0.7 vol.-%. The integration of the local capacity-derived c_v values yields averages $c_{v\,mean}$ ranging from 0.55 to 0.7 vol.-%. The solids concentration profiles exhibit the typical structure with increased solid concentration in the wall zone which has been found by a number of authors in cold model CFBs.

The capacitance probe may also be used as a tool for monitoring the combustor's operating performance. Fig. 7 shows the response of the probe signal to an

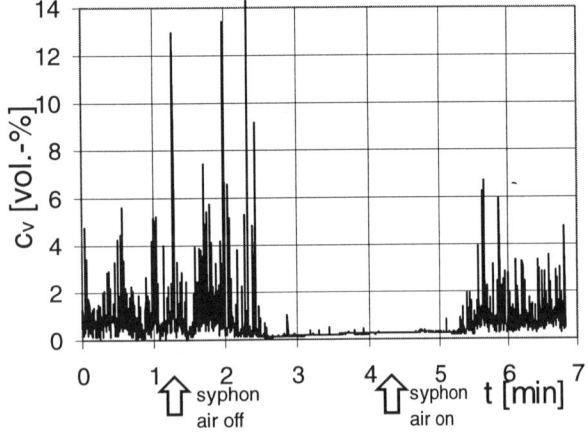

Fig.7 Process observation with the uncooled guarded probe (TUHH CFB combustor, 850°C, h =11.5 m, r = -37 mm).

interruption in the solids external circulation. Shutting off the siphon air interrupts the solids flow which, after one minute leads to a removal of the solids registered by the probe which was located 11.5 m above the distributor level. After turning the siphon air on again, it takes about one minute until the probe registers the filling of the riser. The capacitance probe seems thus to be suited for the diagnosis of disturbances in the local solids flow which, for example, may be caused by variations in the mass flow and quality of the fuel feed or by disturbances in the external solids recycle.

<u>Measurements in the Industrial Scale Reactor</u>

The capacitance probe was introduced through the sampling ports no. 4, 5 and 6 the location of which may be taken from Figure 3. Since it was known from previous measurements by Werdermann [7] at the same combustor, that this unit is characterized by a pronounced temperature profile in the horizontal direction, and since it was known at the same time that the solids volume concentrations calculated from a capacitance probe signal are sensitive against temperature variations [3], the temperature profiles were measured first. The measurements were carried out with a shielded thermocouple (Werdermann and Werther 1994). The resulting temperature profiles are shown in Fig. 8. It is seen that the bulk temperature of 820-850°C is sharply dropping to temperatures between 660 and 710°C close to the wall in a thin layer the thickness of which may be estimated by 200-400 mm. In order to take this temperature variation into account, the dependence of the

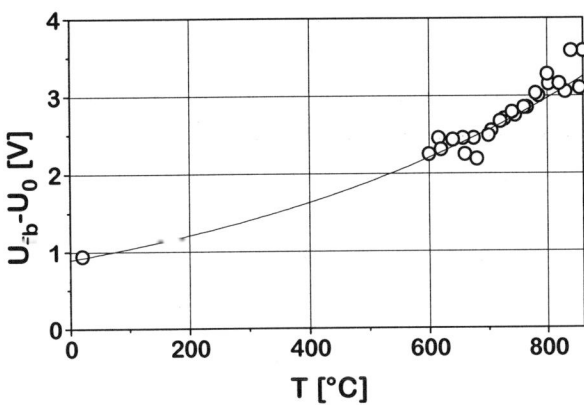

Fig. 9. Difference (U_{fb}-U_0) as a function of the temperature (TUHH bubbling fluidized bed combustor, bed material from Flensburg combustor).

fixed bed voltage on temperature was separately investigated with a bed-ash sample of the Flensburg combustor in the TUHH bubbling fluidized bed combustor. The results are plotted as voltage differences (U_{fb} - U_0) against temperature in Fig. 9. The measurements have been described by a regression curve.

$$U_{fb} - U_0 = 0.89 \cdot e^{0.0015 \cdot T}, \quad (5)$$

where U_{fb} - U_0 is in Volt and the temperature has to be inserted in °C. This regression curve has been used in the treatment of the capacitance probe signals taken in the Flensburg combustor. Figure 10 shows some results. Local time-average solids volume concentrations are plotted against the distance x from the fin of the membrane wall. The accumulation of solids in the wall region at port no. 6 is more pronounced, which may be due to a general accumulation of solids in the corner. The comparison with the geometry of the membrane wall, which is indicated in Figure 5, leads to the conclusion that the accumulation of solids occurs mainly in the space between the neighboring tubes.

The measurements of the local time averaged solids velocities obtained by cross-correlation exhibit completely different horizontal profiles (Figure 11). Immediately at the wall at ports no. 5 and 6, i.e. at a height of 17.3 m downward velocities of nearly 3 m/s are observed whereas at a level of 20.6 m, at point no. 4, the downward velocity at the wall is only 1.5 m/s. With increasing distance from the wall, the solids velocities increase until a mean value of zero at a distance of 80-100 mm from the wall is attained. Thereafter, the

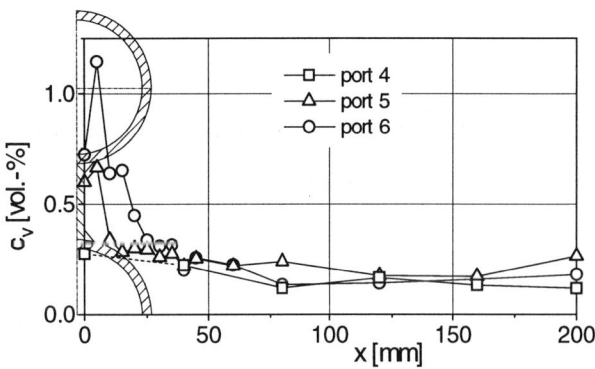

Figure 10 Local particle volume concentration in the Flensburg combustor together with the shape of the membrane wall (x is measured as the distance from the fin; the ports are always located in the middle of the fin; each data point is a mean value over 1 min, data acquired at 100 Hz).

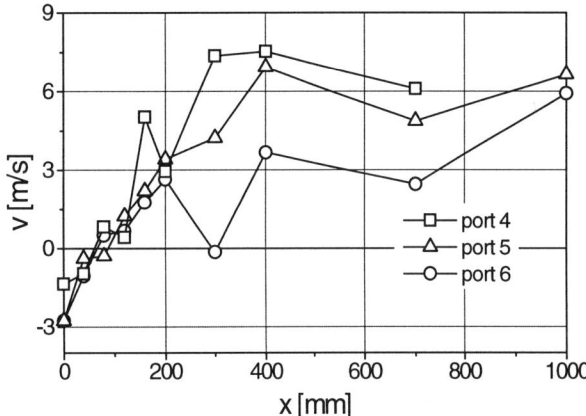

Figure 11 Horizontal distribution of local time-average solids velocities inside the Flensburg combustor.

solids velocities are increasing further, until at a distance of about 400 mm, maximum values between 4 and 7 m/s are reached. It may be noted that the location, where the zero velocity is observed agrees quite well with the boundary of the fluid dynamic wall-zone, defined where the net solids mass flux is zero.

Werdermann did his suction probe measurements [7] in the same combustor at the same locations. He found that the zero net solids flux in the Flensburg combustor occurred at a distance of about 100 mm from the fin of the membrane wall, which is quite close to the 80-100 mm distance observed here for the zero solids velocity.

CONCLUSION

In this paper, a novel capacitance measurement system for the determination of solids volume concentrations and solids velocities at ambient and at elevated temperatures up to 850°C is presented. The measurement system is characterized by an active guard circuitry which keeps the signal drift extremely low. The probes itself are of the needle type for a minimum disturbance of the flow. Both water-cooled and uncooled probes were designed. On the basis of the high signal stability achieved, a calibration of the signals became possible. With a suitable calibration routine and a signal processing for noise correction, it was possible to measure local solids volume concentrations and velocities inside laboratory, semi-technical and industrial-scale fluidized bed combustors. The high signal resolution and stability together with the simple and robust probe setup may offer new possibilities for experimental determination of flow patterns and process control in experimental and industrial reactors with high-temperature gas/solid flows.

ACKNOWLEDGMENT

The authors gratefully acknowledge the help of the staff and the management of the Stadtwerke Flensburg. Special thanks are due to Jörg Rossius, who contributed to the signal calibration.

NOTATION

A	Surface of cross section	[m²]
$c_{v\,fb}$	Fixed bed solids volume concentration	[-]
$c_{v\,\Delta p}$	Solids volume concentration by pressure drop	[-]
c_v	solids volume concentration	[-]
i, j, m, n	Integer numbers in Equation 4	[-]
K	Calibration constant	[1/V]
U	Measured voltage	[V]
U_0	Voltage related to $c_v=0$	[V]
$\overline{\overline{U'}}$	Voltage defined by Equation 4	[V]
U_{fb}	Voltage related to $c_{v\,fb}$	[V]
U'	Voltage defined by Equation 4	[V]
x	penetration depth in riser cf. Figure 3	[m]
y	cf. Figure 3	[m]

LITERATURE CITED

1. **Almstedt, A. E. and E. Olsson,** „Measurement of Bubble Behavior in a Pressurized Fluidized Bed Burning Coal, Using Capacitance Probes" Proc. 7th Int. FBC Conf., Philadelphia, USA, 1982, pp. 89-98.

2. **J. Werther and O. Molerus,** „The Local Structure of Gas Fluidized Beds- I. A Statistically Based Measuring System", Int. J. Multiphase Flow, 1, (1973) 103.

3. **B. Hage, J. Werther, K. Narukawa and S. Mori,** „Capacitance probe measurement technique for local particle volume concentration in circulating fluidized bed combustors" J. Chem. Eng. Japan, 29, 4 (1996) 594-602

4. **C. H. Soong, K. Tuzla and J. Chen,** „Experimental Determination of Cluster Size and Velocity in Circulating Fluidized Beds", Preprints Fluidization VIII, Tours, France, 1995, pp. 1-8.

5. **J. Werther, T. Ogada and C. Philippek,** „N2O emissions from the fluidized bed combustion of sewage sludge", Journal of the Institute of Energy, 6 (1995) 93.

6. **G. Daradimos, H. Hirschfelder, L. Eickenberg, J. Laging und M. Trost,** „Betriebs- und Inbetriebnahmeerfahrungen aus zwei Dampferzeugungsanlagen mit ZWS-Feuerung der Stadtwerke Flensburg GmbH," VGB-Konferenz "Wirbelschichtfeuerung und Dampferzeugung 1988", Essen, V14.

7. **C. Werdermann and J. Werther,** „Heat Transfer in Large Scale Fluidized Beds of Different Sizes", Proc. Int. Conf. Circulating Fluidized Beds IV (A. Avidan, ed.), 1994, pp. 428-435.

Hydrodynamic Aspects of Downflow Gas-Solids Reactors

N.K. Kimm[1], M.A. Forcinito[2], and F. Berruti[3]
[1]BASF Corporation, Mt. Olive, NJ
[2]University of Calgary, Calgary, AB, Canada
[3]University of Saskatchewan, Saskatoon, SA, Canada

The hydrodynamic flow structure of fluid catalytic cracking particles in laboratory scale cocurrent downflow reactors is investigated. The flow characteristics and the typical densification in the radial solids concentration profiles experimentally observed close to the walls are modeled using two different approaches. First, an empirical correlation, incorporating the operating conditions of solids circulation rate and superficial gas velocity is proposed which adequately correlates published data of solids hold-up away from entrance and exit effects (fully developed flow). This correlation describes the suspension homogeneity and hence, degree of gas-solid contact efficiency, according to the operating conditions. Second, the concept of cellular automata is used to simulate the downflow reactor hydrodynamics by envisioning the gas-solids suspension as a set of tractable individual particles. The cellular automaton is based on simple rules governing the particle-particle and particle-wall interactions as well as on the particle velocity profile. The rules for these interactions are derived based on experimental evidence using high speed cinematography. The cellular automata is able to describe the mechanism by which the densification in the radial solids hold-up occurs.

Gas-solids downflow cocurrent circulating fluidized bed reactors (CFB downers) have been proposed in the literature and patents to overcome some of the disadvantages of the more conventional upflow cocurrent circulating fluidized bed reactors (CFB risers). Due to the effects of gravity, risers experience backmixing of solids at the wall under certain operating regimes. In downers, due to both gas and solids traveling downward in the direction of gravity, the flow of solids is generally much more uniform.

In the fluid catalytic cracking (FCC) process, for example, backmixing of solids in risers due to gravity, clustering, and radial solids segregation results in a wide distribution of solids residence times, potentially lowering gasoline yields. Due to the strong tendency for coke formation, the FCC process demands a very short, but uniform, contact time between gas and solids.

---

1 BASF Corporation, Mt. Olive, N.J.
2 University of Calgary, Calgary, AB
3 University of Saskatchewan, Saskatoon, SA

The downer may offer significant operational advantages over the riser for the FCC and other processes. Previous investigations have shown that particle flow in downers is much more uniform radially than in risers and that reduced solids backmixing leads to shorter and more uniform solids residence times (Zhu *et al.* [1]). However, there have been limited attempts to model the hydrodynamics thus far (Bai *et al.* [2,3], Bolkan-Kenny *et al.* [4]).

The radial flow structure of gas-solid suspensions in downers, although generally more uniform than in risers, exhibits some non-uniformities. The solids hold-up, or solid fraction, is relatively flat at the center of the downer but typically shows a peak next to the wall. The location of the maximum in solids concentration has been experimentally determined to be at approximately r/R = 0.94 (Zhu *et al.* [1]). From the maximum to the downer wall, there is a region of low solids hold-up. In risers, on the other hand, this high density region has been found to increase toward and be highest at the wall.

Empirical model

A hydrodynamic model capable of describing the radial flow structure of FCC particles in a small diameter downer at fully-developed flow conditions is proposed. The radial solids hold-up data for a FCC catalyst-air system in a downer with a diameter of 140 mm i.d. and height of 5.8 m were obtained at Tsinghua University (Bai et al. [2], Wang et al. [5]). The steady-state isothermal model incorporates the key operating parameters of solids flux and superficial gas velocity. Inter-particle resistances are neglected considering the rapid mixing of gas and particles and good particle-gas contact efficiency present in the downer. Intra-particle resistance can also be considered negligible because of the small particles sizes.

Two parameters, n and k, have been defined allowing the experimental hold-up profiles to be adequately correlated. The following function based on the hold-up at the center of the downer was found by regression and correlation of all data available (for $0 \leq r/R \leq 0.94$).

$$(1 - \varepsilon) = (1 - \varepsilon)_c + (r/R)^n + k\,(r/R)^4 \quad (1)$$

where

$$(1 - \varepsilon)_c = 9.96 \times 10^{-5}\,U_o + 4.97 \times 10^{-5}\,G_s + 4.39 \times 10^{-3}$$

$$n = 1.31\,U_o - 0.187\,G_s + 82.7$$

$$k = -5.67 \times 10^{-4}\,U_o + 4.54 \times 10^{-6}\,G_s + 6.93 \times 10^{-3}$$

The first term in Equation (1) describes the hold-up in the center of the downer. The second term is the radial distance to a large exponent, which allows the large increase in the hold-up near the wall to be reproduced. The third term uses the exponent of value four to represent the slight curvature in the hold-up increase prior to the maximum, from approximately $r/R = 0.7$ to 0.94. The parameter k reflects the magnitude of this increase.

The parameters n and k were fit to the experimental data using the criteria that the sum of squares between the model calculation and the experimental data should be a minimum and that the values of the maximum hold-up agree within a tolerance of 2%. From the point of maximum hold-up to the wall, the hold-up was assumed to decrease linearly to zero, corresponding to the no-slip condition. Figure 1 presents a comparison of the model with the results of Bai et al. [2] and Wang et al. [5]. The model was found to deviate from the experimental data by a maximum of 7%.

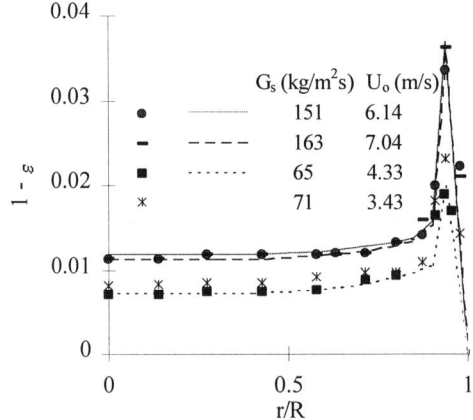

Figure 1. Solids Hold-up in Radial Direction from Model and Data of Bai et al. [2] and Wang et al. [5].

The n parameter is inversely proportional to the height of the maximum hold-up and the k parameter describes the magnitude of the increase prior to the maximum. A combination of the two parameters in the form k/n can then be used to describe the shape of the hold-up profile. A small k/n reflects a flatter curve and a large k/n reflects a greater densification near the wall. From this point onward, $k/n \cdot 10^5$ is referred to as α.

The parameter α increases with increasing solids flux, G_s. As G_s increases, the solids hold-up profiles exhibit a large maximum and the flow is more segregated. This is because the greater amount of particles present result in inelastic collisions, dissipating energy close to the wall. Smaller values of G_s, conversely, indicate a flatter hold-up profile corresponding to more homogeneous flow. In this case, not enough particles are present to result in inelastic collisions.

The parameter α decreases with increasing superficial gas velocity, U_o, since there is less dissipation of energy due to collisions. This can be explained by the change in the velocity profile with increasing U_o. The radial gas velocity profile, and therefore, radial particle velocity profile, is expected to become more uniform with increasing U_o. Therefore, due to the flow being more radially uniform, the distribution of particles over the cross-section of the downer is more uniform and there is a lesser chance for inter-particle collisions. Hence, a lower maximum solids hold- up results. Therefore, the most homogeneous flow and most minimal wall effect is achieved by having a high U_o and low G_s.

Cinematography experiments

The purpose of performing experiments using high speed cinematography was to achieve visual evidence of the inter-particle and particle-wall interactions in the vicinity of the wall. Individual particles were tracked in order to understand the mechanism by which the densification in downers occurs. The interactions revealed through the high speed cinematography could then act as rules providing the basis on which to develop alternative modeling approaches. Experiments were performed by filming uniformly sized polyethylene particles flowing vertically downward along a flat wall in the absence of gas flow using a high-speed video camera. The movements of these particles were tracked in 2 dimensions in order to achieve essentially a 2-dimensional distribution for comparison with the profiles reported in the literature (Bai *et al.* [2], Wang *et al.* [5]).

A net lateral movement away from the wall due to a combination of inter-particle and particle-wall collisions was observed. Furthermore, a decrease in the vertical particle velocity was measured near the wall, even in the absence of a gas velocity gradient. However, the nature of these inter-particle and particle-wall collisions may change once a gas velocity gradient is added. A steeper vertical velocity gradient is expected with gas flow, increasing the opportunity for inter-particle collisions near the wall. Also, it is evident that the types of behavior depend on the physical characteristics of the particles. However, the interactions observed can be used to hypothesize that the densification is a result of both inter-particle and particle-wall interactions coupled with the effects of a particle velocity gradient.

Cellular automata model

A 2-dimensional domain, 0.08 m in width and 0.1 m in length having Cartesian co-ordinates with the origin at the upper left corner is used as the area over which the cellular automaton develops. The domain width of 0.08 m was based on the width of the feeder used for the high speed cinematography experiments, on which the automaton is based. The domain length is much shorter than the length of an actual downer, however, was found to be adequate for the automaton to reach steady state.

A specified number of spherical particles (assumed circular for the 2 dimensional simulation), are introduced at the top of this domain. The distribution of these particles is random and uniform in the horizontal direction. Also, their vertical locations are slightly offset with respect to one another based on a uniform random distribution.

The particles initially fall in the vertical direction with a given vertical velocity. After one time step, a second layer of particles is released from the top of the domain. This continues until all the particles are released. The cellular automaton is allowed to develop over time. Certain information is required for each particle which determines its behaviour at the next time step. This consists of: the direction of the particle, its location, and its y-velocity and x-velocity. This information provides a basis for the rules which propagate the automaton.

It is assumed that the system is 2 dimensional, that the mean free path of a particle is short enough to assume that the particle-wall interactions will be unaffected by the shape of the downer wall, and the circular cross-section of the downer can be approximated by a flat surface. All the particles are assumed to be uniform in size and mass. The particles are assumed to be massive enough so that turbulent velocity fluctuations in the gas phase can be ignored. Finally, inter-particle collisions are assumed to be perfectly elastic and frictionless. There are four rules on which the cellular automaton is based. Rules 1, 2 and 4 are based on the results of the high speed cinematography.

1. Boundary Condition. If the center of the particle is less than a certain distance from the wall, which is a function of the particle radius, the boundary condition is invoked. This moves particles away from the wall at an angle γ_{pw} when the particle is within a certain distance from the wall.

2. Inter-Particle Collisions. A set of inter-particle collision rules moves the particles in certain directions for one time step following an inter-particle collision. The entire set of particles is checked sequentially from particle 1 through n. For the particle of interest, the position at which the collision occurs is calculated by determining the angle between the two particles. A collision section is assigned according to the 8 disk divisions, separated by 45° and there are five possible directions a particle can travel, separated by the angle γ_{pp}. Based on the direction in which the disk is travelling prior to impact and the section where the collision takes place, a direction following the collision is determined. The exiting direction is a function of the entering direction and location of impact.

3. Gas and Particle Velocity Profiles. A y-velocity profile of the form proposed by Patience et al. [6] for risers, but modified for downers is used, based on the empirical results of Bai et al. [2]. The resulting empirically modified expression is:

$$\frac{U_g(r)}{U_o} = \frac{\tau + 2}{\tau}\left[1 - (r/R)^{25\tau}\right] \quad (2)$$

where
$\tau = 1 + 0.1(U_o/U_p)^{1/2}$
$U_p = G_s/\rho_p$

The particle velocity profile was generated by adding the terminal velocity (calculated using the method described by Kunii and Levenspiel [7]) to the gas velocity profile.

4. Decay in X-Velocity between Time Steps. There is a probability that the x-velocity will decrease after one time step. This probability is larger, the greater the magnitude of the x-velocity, due to the influence of gravity.

The following parameters were tested in order to determine their effects on the solids hold-up profile: the angle after particle-particle collisions (γ_{pp}), the angle after particle-wall collisions (γ_{pw}), superficial gas velocity, solids flux, degree of slip, particle diameter, particle density, and boundary condition. The base case was as follows $\gamma_{pp} = 5°$, $\gamma_{pw} = 10°$, $U_o = 10$ m/s, $G_s = 137$ kg/m^{2}s, $d_p = 59$ μm, $\rho_p = 1540$ kg/m^3 and a boundary condition corresponding to a distance equal to 10 times the particle radius.

A population of 1000 particles was used for the simulation. Increasing the particle number did not significantly change the results, and approximately the same densification relative to the mean hold-up was observed for populations greater than 1000. Five layers of particles, consisting of 200 particles per layer were introduced per time step.

Figure 2 shows the solids hold-up versus x distance generated by the cellular automaton for the base case. At a location of x/X = 0.1, there is a maximum in solids concentration (1.3%). The mean, μ, over the x direction is 0.8% and the standard deviation, σ, is 0.15%. The maximum is statistically significant and is 55% larger and more than two standard deviations above the mean. When the U_o is raised to 20 m/s, the maximum solids hold-up (1.4%) is 50% larger than the mean of 0.9%. For a $G_s = 79.5$ kg/m^{2}s, the solids hold-up maximum of 0.65% is 29% above the mean of 0.5%. Hence, the trends in the cellular automaton are consistent with the experimental findings of Bai et al. [2] and Wang et al. [5].

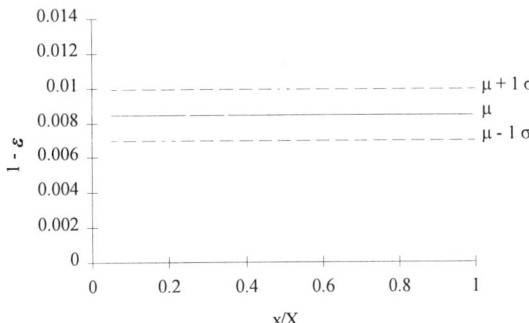

Figure 2. Solids Hold-up in the X-Direction from Cellular Automaton for $G_s = 137$ kg/m^{2}s and $U_o = 10$ m/s.

When the empirical coefficient in Equation 2, indicating the degree of slip at the wall, is reduced from 25 to 10, the maximum solids hold-up is unchanged at $x/X = 0.1$, but is 9% higher at $x/X = 0.05$. When this coefficient is changed to 1, corresponding to no slip at the wall, the maximum is 105% greater than the mean. Changing the angle for particle-particle collisions, γ_{pp}, from 5° to 2.5° results in a maximum solids hold-up of 1.2%, which is only 45% above the mean, compared to 55%, in the base case. This is because the effect of the inter-particle collisions has been minimised. This is because the particles have less x-momentum and are less likely to collide. Changing the angle for particle-wall collisions, γ_{pw}, from 10° to 20°, has little effect on the solids hold-up profile. This means that the angle at which the particles rebound from the wall is likely not solely responsible for the densification mechanism. The densification is a combination of the particles rebounding from the wall and colliding with adjacent vertically travelling particles. Doubling the particle diameter from 59 μm to 118 μm results in a densification of 47%, since there were fewer collisions across the entire x direction.

Conclusion

An empirical, hydrodynamic model is proposed which describes the fully developed radial flow structure of fluid catalytic cracking particles in laboratory scale downers. The study reveals the importance of friction between the gas-solids suspension and the wall in reducing gas-solids contact in laboratory scale downflow reactors. Wall effects are described in terms of changing operating conditions by the model parameter α. High speed cinematography provided visual evidence of inter-particle and particle-wall interactions. The observations made during these experiments provided certain rules of interaction which were incorporated into a cellular automata model. An examination of the parameters shows that the flow structure is particularly influenced by the particle velocity profile, the degree of slip, and the angle after inter-particle collisions, γ_{pp}. The results, however, are meant to show only the qualitative behaviour and do not predict the results found in experimental studies on downflow reactors. Further study would be required to correlate the parameters with the operating conditions. This work does, however, demonstrate that the concept of cellular automata can indeed be used to describe the physical mechanism involved in creating the solids densification in downflow reactors.

Nomenclature

d_p	=	Particle diameter, m
G_s	=	Solids flux, kg/m²s
k	=	Term in Equation (2)
n	=	Exponent in Equation (2)
r	=	Radial coordinate, m
R	=	Downer radius, m
U_o	=	Superficial gas velocity, m/s
U_g	=	Local gas velocity, m/s
U_p	=	Local particle velocity, m/s
x, X	=	Horizontal distance, m

Greek Letters

α	=	model parameter, $k/n \times 10^5$
γ_{pp}, γ_{pw}	=	Angle of particle movement with respect to the vertical following inter-particle and particle-wall collisions
ε	=	Voidage
$(1-\varepsilon)$	=	Solids hold-up
$(1-\varepsilon)_c$	=	Solids hold-up at center of Downer
ρ_p	=	Particle density, kg/m³
τ	=	Term in Equation (2)

Literature cited

1. Zhu, J.-X., Yu Z.-Q., Jin, Y., Grace, J.R. and Issangya, A., "Cocurrent Downflow Circulating Fluidized Bed (Downer) Reactors", *Can. J. Chem. Eng.* 73, October, 662-677, (1995).

2. Bai, D.-R., Jin, Y., Yu, Z.-Q. and Gan, N.-J., "Radial Profiles of Local Solid Concentration and Velocity in a Concurrent Downflow Fast Fluidized Bed", in *CirculatingFluidized Bed Technology III*, eds. P. Basu, M. Horio and M. Hasatani, Pergamon Press, Toronto, 157-162, (1991a).

3. Bai, D.-R., Jin, Y., Yu, Z.-Q. and Gan, N.-J., "Gas-Solids Flow Patterns in a Concurrent Downflow Fast Fluidized Bed (CDCFB)", *J. Chem. Ind. & Eng. of China* (English Edition), 6(2), 171-181, (1991b).

4. Bolkan-Kenny, Y.G., Pugsley, T.S. and Berruti, F., "Computer Simulation of the Performance of Fluid Catalytic Cracking Risers and Downers", *Ind. Eng. Chem. Res.* 33, 3043-3052, (1994).

5. Wang, Z., Bai D. and Jin, Y., "Hydrodynamics of Cocurrent Downflow Circulating Fluidized Bed (CDCFB)", *Powd. Technol.*, 70, 271-275, (1992).

6. Patience, G.S., Chaouki, J. and F. Berruti, "Gas Phase Hydrodynamics in Circulating Fluidized Bed Risers", in *Multiphase Reactor and Polymerization System Hydrodynamics - Advances in Engineering Fluid Mechanics Series*, ed. N.P. Cheremisinoff, Gulf Publishing, 255-296 (1996).

7. Kunii, D., and Levenspiel, O., *Fluidization Engineering*, ed. H. Brenner, 2nd Ed., Series in Chemical Engineering, Butterworth-Heinemann, Toronto, (1991).

A Modern Development of Friction Factors in Porous Media by Analogy to Turbulence

E. A. Stephan, M. S. Willis, R. Vengimalla, and V. K. N. Yerra
Department of Chemical Engineering, The University of Akron, Akron, OH 44325

The friction factor correlation for flow in porous media is based upon experiments conducted in the early 1900's. The explanations for the correlation have focused on a heuristic, macroscopic approach. In this paper, a fundamentally based derivation of the friction factor for porous media flow is developed. This methodology elucidates the flow effects at the pore scale and identifies the effective surface area as the most important solid phase material property for flow evaluation. A novel procedure is provided for the indirect measurement of the effective specific surface area.

The Ergun Equation has been used in the estimation of equipment sizes, flow rates, and pressure drops in packed beds since the early 1900's [1]. A fundamental review of this correlation has not been conducted since several empirical advances were made in the 50's and 60's. The goal of this work is two-fold: (1) by using averaging, scale, and the basic concepts, to develop a friction factor correlation, and (2) to illustrate the use of the Ergun Equation, with the necessary modifications, to determine the effective specific surface area.

Based on the mean value theorem, averaging a given function will produce a unique value. A given average value, however, may apply to an infinite number of functional forms. This allows for the discretionary choice of a starting function such that once the averaged functions have been determined, they can apply to any number of equivalent systems.

In turbulent flow, the concept of time averaging is used to eliminate unknown information about the velocity profiles occurring within an eddy. The time averaging introduces a second constitutive relation, the Reynolds stress equation, to account for this information. In this work, the unknown geometry of the porous media is spatially averaged to obtain the averaged governing equations. As shown later, averaging the mass and momentum balances introduces the porosity and available surface area, respectively.

MODEL ONE

To represent a porous media, a tube highly packed with small particles is chosen, shown in Figure 1(a). The particle surface area available for momentum transfer is very large and it is assumed that there is no momentum transfer between the fluid and the tube wall. The system can be represented equivalently by a porous media compressed against the wall, without any loss of generality, shown in Figure 1(b). It is further assumed the system is at steady state, the fluid and solid phases have constant density, the solid phase is stationary, and there are no fluid-fluid interactions. The flow pattern within the media is at the turbulent, time-averaged scale and the

Department of Chemical Engineering
The University of Akron, Akron, OH 44325-3901

momentum balance will be comprised of viscous and pressure forces.

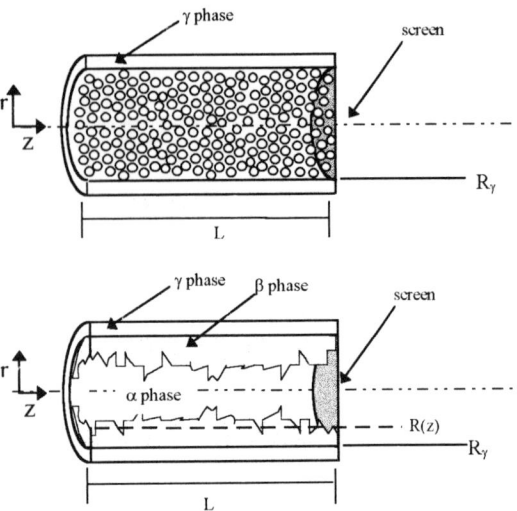

Figure 1. Representative Model (a) general (b) consolidated

This system can be used to determine the average functional values of the basic concepts. These average values can then represent any number of equivalent systems. This process is illustrated in Figure 2.

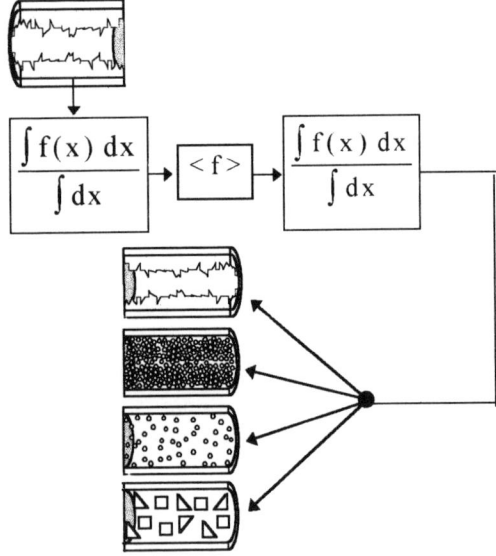

Figure 2. Concept of Spatial Averaging

Governing Equations - Model One

Spatial averaging of the continuum scale mass balance [2,3] yields the mezzo-scale continuity balance and introduces the porosity as a new material property.

$$\frac{d}{dz}\left[\varepsilon^\alpha \langle \upsilon_z^\alpha \rangle\right] = 0 \qquad (1)$$

The fluid phase porosity is defined as the ratio of porous media cross section to tube cross section. The fluid phase and solid phase porosity are given in Equations (2) and (3). Note that the fluid phase porosity is the fractional area available for flow.

$$\varepsilon^\alpha = \frac{\pi R_{(z)}^2}{\pi R_\gamma^2} \qquad (2)$$

$$\varepsilon^\beta = \frac{\pi \left(R_\gamma^2 - R_{(z)}^2\right)}{\pi R_\gamma^2} \qquad (3)$$

Spatial averaging the fluid continuum momentum balance yields the mezzo-scale momentum balance.

$$a^{\beta\alpha} \tau_{rz}^\alpha \Big|_{pore} + \varepsilon^\alpha \frac{dP^\alpha}{dz} = 0 \qquad (4)$$

Averaging leads to a second material property, the particle surface area available for momentum transfer, $a^{\beta\alpha}$. It is the ratio of porous media circumference in contact with the fluid to the tube cross section.

$$a^{\beta\alpha} = \frac{2\pi R_{(z)}}{\pi R_\gamma^2} \qquad (5)$$

Law of Dynamic Similarity - Model One

The definition of the friction factor is the total drag divided by the characteristic area times the kinetic energy [2].

$$f = \frac{\int_0^L \int_0^{2\pi} \tau_{rz}^{\alpha}\big|_{R_{(z)}} R_{(z)} \, d\theta \, dz}{(2\pi R_{(z)} L)\left[\frac{1}{2} \rho^{\alpha} \langle v_z^{\alpha}\rangle_0^{\alpha^2}\right]} \quad (6)$$

In dimensionless form, Equation (6) becomes Equation (7). For a known velocity profile and constant porosity, the friction factor is a function of only the Reynolds number. Thus, the Law of Dynamic Similarity applies to porous media flow.

$$f = \frac{2\varepsilon^{\alpha} \int_0^{a^{\beta\alpha} L}\left(-\partial v_z^*/\partial r^*\right)_{a^{\beta\alpha} R_{(z)}} dz^*}{\left(a^{\beta\alpha} L\right) Re} \quad (7)$$

The Reynolds number, obtained by dimensional analysis of Equation (4), uses $a^{\beta\alpha}$ as the characteristic length. It is the ratio of drag force to pressure force, analogous to single phase fluid tube flow [2,3].

$$Re = \frac{\varepsilon^{\alpha} \langle v_z^{\alpha}\rangle_0^{\alpha} \rho^{\alpha}}{\mu \, a^{\beta\alpha}} \quad (8)$$

The friction factor can be written in experimental quantities, shown below.

$$f = \frac{2\pi R_{\gamma}^4 \left(\varepsilon^{\alpha}\right)^3 (P_0 - P_L)}{a^{\beta\alpha} L \rho^{\alpha} Q^2} \quad (9)$$

The relations shown apply to systems where no momentum is transferred to the wall. As the Reynolds number increases, the surface area available for momentum transfer decreases and the current model no longer applies.

GENERALIZED MODEL

A generalized model is required which satisfies the limiting cases of high and low surface areas. Figure 1 represents the case of high surface area described by Model One. The case of low surface area can be represented by the systems shown in Figure 3. The same assumptions apply, except fluid momentum is transferred to both the particles and tube wall.

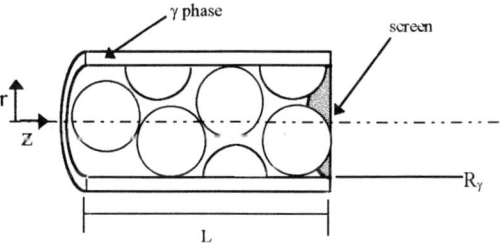

Figure 3. Generalized Model, Low Surface Area

As before, the porous media can be consolidated against the tube wall. The material can be further concentrated to one end of the tube, shown in Figure 4, due to the summation property of integration.

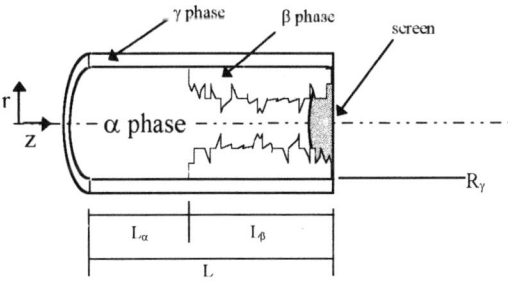

Figure 4. Model Incorporating Momentum Transfer to Wall

Governing Equations - Generalized Model

Analogous to Model One, the continuum scale mass and momentum balances are spatially averaged to obtain the mezzo-scale equations. The mass balance derivation yields results identical to Equation (1). The momentum equation must be averaged over two areas: one of direct momentum transfer to the wall and one with momentum transfer to the particles, yielding the mezzo-scale momentum balance. This averaging process introduces a third material property, the surface area available for momentum transfer from the fluid to the tube wall.

$$a^{\gamma\alpha} = \frac{2\pi R_\gamma}{\pi R_\gamma^2} = \frac{2}{R_\gamma} = \frac{4}{D_\gamma} \qquad (10)$$

Law of Dynamic Similarity - Generalized Model

By expanding the definition for the Reynolds number and friction factor given previously and applying dimensional analysis, the following relations can be obtained.

$$Re = \frac{\varepsilon^\alpha \langle v_z^\alpha \rangle_0^\alpha \rho^\alpha}{\mu \left(a^{\beta\alpha} + a^{\gamma\alpha}\right)} \qquad (11)$$

$$f = \frac{2\left(P_0 - \varepsilon^\alpha P_L + (\varepsilon^\alpha - 1) P_{L_\alpha}\right)}{\left(a^{\beta\alpha} L_\beta + a^{\gamma\alpha} L_\alpha\right) \rho^\alpha \langle v_z^\alpha \rangle_0^{\alpha 2}} \qquad (12)$$

The friction factor can be analyzed for two limiting cases. For a high surface area media, Equation (12) will simplify to the friction factor given previously in Equation (9). For the case of low surface area, the limit of the friction factor is the Fanning equation for tube flow.

$$f = \frac{D_\gamma (P_0 - P_L)}{4 L \left[\frac{1}{2}\rho^\alpha \langle v_z^\alpha \rangle_0^{\alpha 2}\right]} \qquad (13)$$

ANALYSIS OF THE ERGUN EQUATION

The equations for the friction factor and Reynolds number presented can be related to the definitions used by Ergun [1]. Shown in Figure 5, the high and low surface areas can be represented by the limiting regions of the Ergun Equation. This plot can be used to determine the specific surface area of the porous media [2].

By eliminating the surface area between the friction factor and Reynolds number for the case of high surface area, a line given by Equation (14) can be drawn, with a slope of one, from known experimental quantities. The porosity is assumed to be known and of a constant value to satisfy the Law of Dynamic Similarity.

$$f = \left(\frac{2\mu (P_L - P_0)}{L (\rho^\alpha)^2 \langle v_z^\alpha \rangle_0^{\alpha 3}}\right) Re \qquad (14)$$

The intersection of Equation (14) and the Ergun Equation gives the friction factor-Reynolds number combination for the system of interest. The available surface area can then be found numerically.

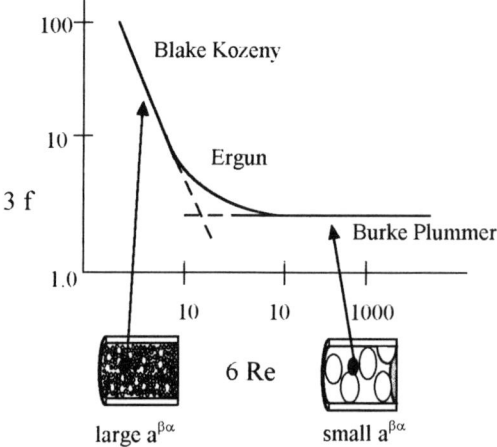

Figure 5. Modified Friction Factor-Reynolds Number Plot

CONCLUSIONS

The fundamental approach utilized defines both a new mechanism and friction factor-Reynolds number relationship to explain the functional form of the Ergun Equation. This is the first practical method to indirectly obtain the effective surface area for a given porous media.

NOTATION

a	effective surface area $[=]$ m^{-2}
D	diameter $[=]$ m
f	friction factor
L	lenght $[=]$ m
P	pressure $[=]$ Pa
Q	flowrate $[=]$ m^3/ sec
R	radius $[=]$ m
Re	Reynolds Number

Greek Symbols

ε	porosity
μ	viscosity $[=]$ kg / m*sec
ρ	density $[=]$ kg / m^3
τ	shear stress
υ	velocity $[=]$ m / sec

Subscripts / Superscripts

α	fluid phase
β	solid phase
γ	containment wall

LITERATURE CITED

1. **Ergun, S.**, *Chem. Eng. Prog.*, **43**, 89 (1952).

2. **Bird, R. B., W. E. Steward, and E. N. Lightfoot**, *Transport Phenomena*, J. Wiley & Sons, New York (1960).

3. **Willis, M. S., et. al.**, "Dispersed Multiphase Theory" in *Advances in Porous Media*, **1**, Corapcioglu (ed.), Elsevier Science Publishers, Amsterdam (1991).

A Volume-Average Scale Model for Fines Migration in Sandstone

E. A. Stephan and G. G. Chase
Department of Chemical Engineering, The University of Akron, Akron, OH 44325

The permeability reduction that occurs in sandstone reservoirs during the water injection phase of secondary oil recovery is a well documented problem in the petroleum industry. Clay minerals naturally found in the sandstone may release if the formation is subject to a water shock, or fine particles may be introduced in the injection water. The fines migrate and are eventually captured within the sandstone pores by a variety of deep bed filtration mechanisms. A literature review reveals that past research is mainly focused at either the macroscopic or at the microscopic level. The macroscopic research does not reveal the mechanisms which influence the permeability reduction, while the microscopic efforts have limited success with scale-up to yield accurate process predictions.

In this work, the fundamental equations and constitutive relations are developed at the volume-average scale, the intermediate between the micro-scale and macro-scale. One-dimensional flow equations are rigorously applied to the process of water injection through a sandstone formation. Then, a macro-scale numerical model is developed for the permeability reduction, incorporating the mechanisms revealed at the volume-average scale. Three cases are considered: (1) the formation is subject to a water shock, (2) the injection water contains fine particles but does not cause a water shock, and (3) a combination of the water shock and injected fines. A parametric study is conducted to determine the parameters critical for future analysis.

During secondary oil recovery, water injection can cause reduction in sandstone permeability. The injected water may contain fine particles or may cause the release of formation fines by a water shock. A water shock is created by the injection water having a different composition from the formation brine. The entrained fines migrate through the bed and are captured by various deep bed mechanisms. Over the past twenty years, a great quantity of research has been conducted in this area. The majority of the effort has focused on the macroscopic or microscopic level, with limited success in developing a correlation incorporating the two scales.

The past work conducted at the microscopic level has greatly enhanced the understanding the mechanisms behind general deep bed filtration. Due to their complexity, however, the models have been limited to simple geometries and most have not been verified by experiment. Macroscopic research has yielded experimental information concerning the specific process of permeability reduction in sandstone during oil recovery, with little investigation of the mechanisms which cause the behavior. The incorporation of the mechanisms controls the usefulness of a correlation. If too many experimental factors are used to impose a "good fit", the predictability is lost while if an extreme theoretical approach is taken, the correlation becomes mathematically untreatable [1]. This work is unique in that the focus is on the volume-average scale. The volume-average equations are the basic concepts applied to a multiphase system, analogous to the basic concepts for a single phase system, but they have "excess terms" to account for phase interactions [2]. The volume-average scale utilizes the microscopic theories to make predictions about the mechanisms which control the observable, macroscopic behavior.

The authors realize that the overall scope of deep bed filtration within a sandstone formation is very dynamic and complex. Therefore, the system is simplified by the following assumptions to allow for preliminary analysis.

- Flow is laminar, isothermal, unidirectional, constant
- The formation is formed of incompressible Berea sandstone with 19% porosity and 8% by weight of kaolinite clay fines.
- The sandstone and fines diameters are considered uniform and constant

Department of Chemical Engineering,
The University of Akron, Akron, OH 44325-3906

- Solid phase density, liquid phase density and viscosity are constant.
- Solid phase sandstone is stationary.
- Entrained fines move with liquid phase velocity.

The "excess term" is used to account for the change in phase of the clay fines. The capture and release are modeled as first order processes, depending on the concentration of the fines within the appropriate phases. Table 1 summarizes the volume-average equations. Equations [1] - [3] represent the solid phase mass, solid phase species, and liquid phase species balances respectively, simplified from the basic concepts. The equation for the excess term is given by Equation [4]. Equation [5] is the relation for pressure drop determined from the Blake-Kozeny expression and Equation [6] is Darcy's Law for permeability. The equations are applied to a preliminary model to determine the significant parameters and areas of required experimentation. The model will eventually be utilized to determine the effects of pre-filtration of injection water on formation permeability loss.

Table 1. Summary of model equations.

$\rho^s \frac{\partial \varepsilon^s}{\partial t} - E_f^L = 0$	[1]
$\rho^s \varepsilon^s \frac{\partial \omega_f^s}{\partial t} + E_f^L (\omega_f^s - 1) = 0$	[2]
$\rho^L \frac{\partial (\varepsilon^L \omega_f^L)}{\partial t} + \rho^L \varepsilon^L \upsilon_z^L \frac{\partial \omega_f^L}{\partial z} + E_f^L = 0$	[3]
$E_f^L = \beta \rho^L \varepsilon^L \omega_f^L - \alpha \rho^s \varepsilon^s \omega_f^s$	[4]
$\Delta P = \frac{B \mu L Q (\varepsilon^s)^2 (a_s)}{g_c A (\varepsilon^L)^3}$	[5]
$k = -\frac{Q \mu L}{\Delta P \, A}$	[6]

MODEL DEVELOPMENT

Three cases are modeled: (1) the formation subjected to a water shock, (2) the formation subjected to fines entering with the injected liquid, and (3) the combination of water shock and injected fines. To simulate a water shock, the core is assumed to be initially saturated with NaCl solution. At a specified time, fresh water is pumped into the core. For the case of injected fines, the fines are assumed to be the same size and composition of those initially in the bed.

Past studies [3-5] have shown the release coefficient, α, is a function of the salt concentration, temperature, and flowrate of the injected water. Under isothermal, constant flow conditions, the clay particles release once the salt concentration in the bed has fallen below a critical level. The change in salt concentration is modeled as a tank-in-series [6]. The critical salt concentration (CSC) depends upon the specific salt-clay ion combination. For the current model, NaCl will be used as the entrance fluid and the CSC for NaCl has been shown to be 0.072 M. It has also been proposed that only a small amount of the total fines present in the bed will release. Initial experiments place this value at a mass fraction of approximately 0.076 [3].

The capture coefficient, β, has been shown to be a function of flowrate and the type of capture mechanism [3]. The fines are assumed to be captured by direct interception, with a direct linear dependency upon the amount of fines present in the liquid phase.

The pressure drop is modeled using a modified Blake-Kozeny expression, Equation [5]. While the Blake-Kozeny equation is valid for porosities in the 0.35 to 0.55 range for laminar flow, it will provide an estimate for the current system. Further work will require a more accurate equation to model the pressure drop. The surface area within the Blake-Kozeny expression, a_s, is unknown and must be determined from experimental data. For initial modeling, it is estimated using the surface to volume ratio and incorporating sphericity [7]. The permeability is calculated by Darcy's Law, Equation [6].

MODEL RESULTS

A parametric study was conducted on the three cases to determine which factors require further experimental study. First, mathematical convergence of the time and distance steps for the model was investigated. The distance steps showed convergence at 0.127 cm. Past experiments have recommended a distance step of 0.5 cm based upon experimentation [3], but the larger step clearly results in a drastically difference prediction. The time step size was not critical for convergence, so one second was chosen.

Water Shock with No Entrance Fines

As shown in Figure 1, a changes of as much as 50% in either the release or capture coefficients caused a change in the permeability of less than 10%. This shows an assumption of constant coefficients for a given flowrate is valid. The model was simulated at a mass fraction of releasable fines equal to 0.06, with baseline conditions of $\alpha = 0.038$/second and $\beta = 0.024$/second.

The most sensitive parameter for water sensitivity prediction appears to be the mass fraction of releasable fines, shown in Figure 2. An 8% change in the mass fraction created a 35% change in the predicted permeability, while a 50% change in mass fraction caused a permeability change of greater than 60%. The release and capture coefficients were set equal to the baseline conditions given earlier.

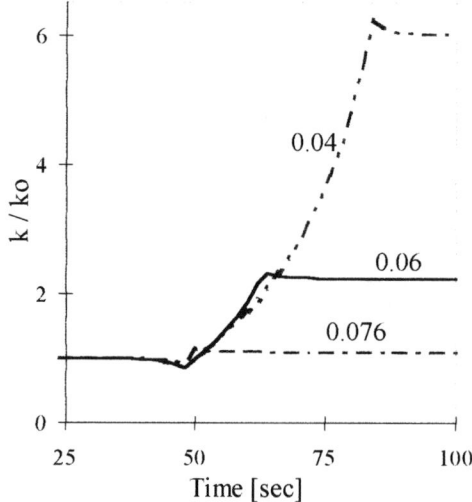

Figure 2. Effect of mass fraction of releasable fines.

Entrance Fines - No Water Shock

If fines are entrained in the injection liquid and no water shock occurs, the model is effectively one depicting pure filtration, shown in Figure 3. As expected, the greater the mass fraction of fines injected, the greater the drop in permeability. However, the model lacks the distinctive sharp drop in permeability shown by experimental results given in literature [3-5]. This discrepancy arises perhaps due to the capture coefficient value or the functional dependencies chosen. Past studies have also limited the fines migration to individual sections, while the current model allows for fines migration through the entire bed.

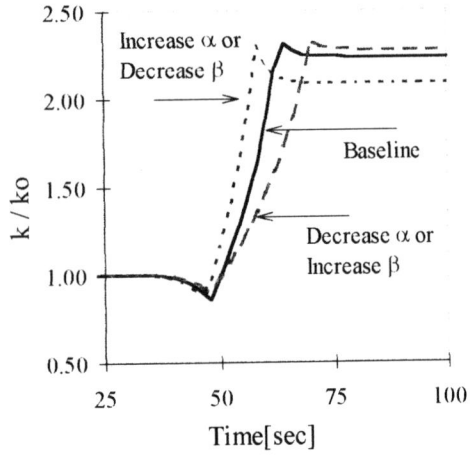

Figure 1. Effect of release and capture coefficients.

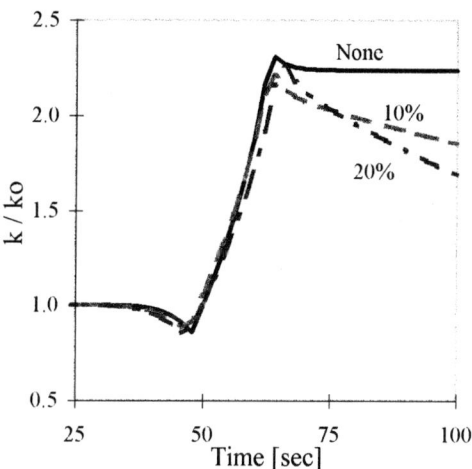

Figure 3. Effect of fines injected into bed.

Combination Injected Fines / Water Shock

The results for the combined case are similar to the first case, with the release and capture coefficient values having little effect on the permeability change. Figures 4 and 5 illustrate the results of simultaneous release and capture of fines. In Figure 4, the mass fractions of injected fines was varied at a releasable mass fraction of fines equal to 0.06. Figure 5 further demonstrates the great sensitivity exhibited by the model to changes in the mass fraction of releasable fines, for an injection mass fraction of 0.1.

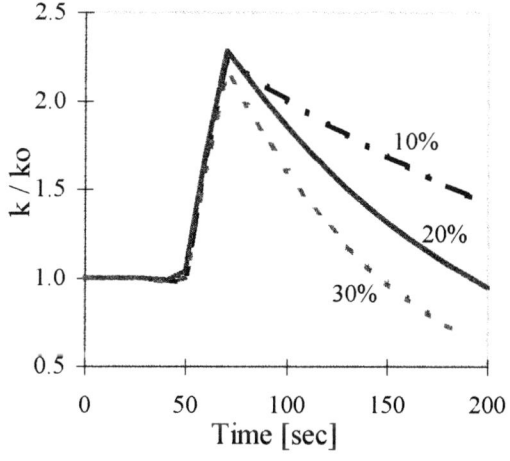

Figure 4. Release & capture, with varying injected fines.

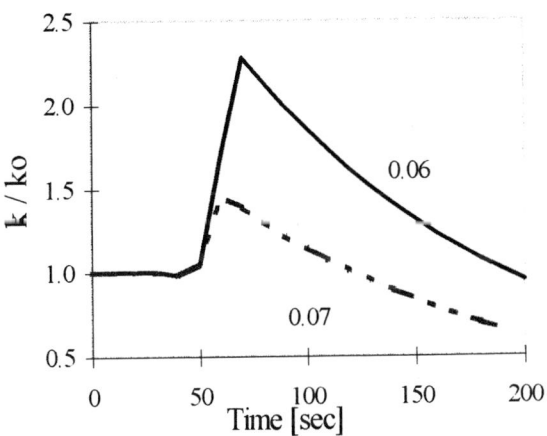

Figure 5. Sensitivity of mass fraction of releasable fines.

CONCLUSIONS AND FUTURE WORK

The results of the parametric study on the current model indicate the most sensitive factor affecting the permeability is the mass fraction of releasable fines. Very little published experimental data exists on this parameter. Due to the highly sensitive nature of the model to this parameter, further experimental studies are required to verify an accurate value.

The capture coefficient was taken as a constant, with only the capture mechanism of direct interception considered. In future work, this coefficient should change magnitude based upon the type of dominant mechanism. It should also vary dependent upon the size, type, and distribution of fines injected into or released from the bed.

ACKNOWLEDGEMENT

This work has been gratefully supported by the Produced Water Conference Group in Houston, Texas and The University of Akron.

NOTATION

a	surface area [=] cm^{-2}
A	area of bed [=] cm^2
B	constant, Blake-Kozeny equation
E	excess term [=] $kg/cm^3 *sec$
g_c	gravitational constant [=] $kg*cm/N*sec^2$
k	permeability [=] darcys
L	length of bed [=] cm
ΔP	pressure drop [=] Pa
Q	flowrate [=] cm^3/sec

Greek Symbols

α	release coefficient [=] sec^{-1}
β	capture coefficient [=] sec^{-1}
ε	porosity
ω	mass fraction
ρ	density [=] kg/cm^3
μ	viscosity [=] $kg/cm*sec$
υ	velocity [=] cm/sec

Subscripts / Superscripts

f	fine particles
l	liquid phase
s	solid phase

LITERATURE CITED

1. **van Deemter, J. J.**, "Basics of Process Modeling", *Chem. Engng. Sci.*, **37**, 657 (1982).
2. **Chase, G. G. and Willis, M. S.**, "Compressive Cake Filtration", *Chem. Engng Sci.*, **47**, 1373 (1992).
3. **Khilar, K. C., and H. S. Fogler**, "Water Sensitivity of Sandstones", *SPEJ*, **55** (Feb., 1983).
4. **Khilar, K. C., and H. S. Fogler**, "The Existence of a Critical Salt Concentration for Particle Release", *J. Colloid. Interface Sci.*, **101**, 214 (1984).
5. **Kia, S. F., et. al.**, "Effect of Salt Composition on Clay Release in Berea Sandstone", *SPEPE*, 277 (1987).
6. **Levenspiel, O.**, *Chemical Reaction Engineering*, J. Wiley and Sons, New York, 290 (1972).
7. **McCabe, W. L., et. al.**, *Unit Operations for Chemical Engineers*, McGraw-Hill, New York (1985).

Simultaneous Sulfur and Metal Capture by Lime During Fluidized Bed Combustion

T. C. Ho, C. C. Shie, K. S. Wang, and J. R. Hopper
Department of Chemical Engineering, Lamar University, Beaumont, TX 77710

The objective of this work was to investigate the effect of sulfur on metal captured by lime during fluidized bed combustion. Experiments were carried out in a well-instrumented 76 mm ID fluidized bed combustor burning artificially prepared test material in a bed of lime. The test material was prepared from wood pellets spiked with various concentrations of metals and sulfur. The metals involved in the test were compounds of lead, cadmium and chromium. The effect of sulfur on the metal capture process was also theoretically investigated through performing equilibrium calculations based on the minimization of system free energy. The observed results indicated that lime is capable of capturing both metals and sulfur simultaneously during fluidized bed combustion. The existence of sulfur may either enhance or inhibit metal capture depending on combustion temperature. Air flow rates did not appear to affect the capture efficiency as long as the bed is under good fluidization operations.

Many industrial wastes and wastes from Superfund sites are contaminated with various hazardous organic constituents. An effective method for treating these wastes is through thermal oxidation. The treatment, however, would inevitably produce air pollutants, such as SO_x, NO_x, and HCl. Toxic metal fumes may also be generated if the wastes also contain metals. While the health and environmental effects associated with the traditional air pollutants, i.e., SO_x, NO_x, and HCl, have been well-recognized, the U.S. EPA has reported that metals can account for almost all of the identified cancer risks from waste incineration systems [1].

Metals will not be destroyed during thermal treatment of metal containing waste. They will either volatilize or stay in burned ash. The volatilized metals will recondense during the cooling of the flue gas. The condensation process either will form metal fumes through homogeneous condensation or deposit on existing surfaces through heterogeneous condensation. The formed metal fumes are generally sub-micron in size which have great potential to escape from the treatment system and create metal emission problems.

Current practice for metal emission control during waste incineration employs traditional air pollution control devices (APCDs), such as venturi scrubbers, electrostatic precipitators (ESPs), baghouses and spray dryer scrubbers, to collect metal fumes at the cold-end of the incineration process [2]. These current APCDs, however, may not always control metal emissions effectively and consistently because the size of the metal fumes to be collected is extremely small [1].

An effective technology for metal emission control during waste incineration is to use sorbents to capture metals during high temperature combustion, especially under fluidization operation. Our recent publications have clearly indicated that metal capture efficiency can be as high as 97% depending on metal species, sorbent involved and operating parameters [3,4,5]. The previous studies, however, have not investigated the effects of other pollutants on the efficiency of metal capture by sorbents.

This study investigated both experimentally and theoretically the interacting effects of sulfur on the capture of various metals by lime during fluidized bed combustion. In addition, the potential of employing the sorbent for simultaneous metals and sulfur capture was also investigated.

THEORETICAL

Chemical absorption reactions between metal vapors and a variety of sorbents at high temperatures have been observed both in packed beds and fluidized beds [3-9]. Kaolinite and bauxite have been reported to be suitable sorbents for the removal of alkali vapors and lead compounds through chemical absorption [6,7,8,9]. Calcined limestone has also been reported to be an effective sorbent for metal capture [3,4,5]. Most of the captured metals have been observed to be non-leachable through this chemical absorption mechanism. The following reactions between metals and sorbent constituents have been reported [5,7,8,9]:

$$2\ PbO + SiO_2 \longrightarrow Pb_2SiO_4(s) \quad (A)$$

$$CdO + SiO_2 \longrightarrow CdSiO_3(s) \quad (B)$$

$$CdO + Al_2O_3 \longrightarrow CdAl_2O_4(s) \quad (C)$$

$$PbCl_2 + Al_2O_3:2SiO_2 + H_2O \longrightarrow$$
$$PbO:Al_2O_3:2SiO_2(s) + 2\ HCl(g) \quad (D)$$

$$CdCl_2 + Al_2O_3 + H_2O \longrightarrow CdAl_2O_4(s)$$
$$+ 2\ HCl(g) \quad (E)$$

The previous investigations, however, have not examined the interacting effects of other air pollutants, such as SO_x and HCl, on the effectiveness of metal capture by sorbents. In addition to the above metal-sorbent reactions, potential competing reactions among these air pollutants, metals, and sorbents may include the formation of $PbSO_4$, $PbCl_4/PbCl_2$, $CdCl_2$, $CdSO_4$, $CaSO_4$ and $CaCl_2$.

EXPERIMENTAL

Experiments were carried out in a well-instrumented 76 mm ID fluidized bed combustion system. A schematic diagram of the system is shown in Figure 1. The system includes a fluidized bed preheater, a fluidized bed combustor, two cyclones, a wet scrubber, and a baghouse.

In an experimental run, a pre-designated amount of lime (normally 300-600 g) was charged into the incinerator. The system was then preheated and adjusted to the designed temperature (600, 750, or 900°C) and operating condition (normally 3 U_{mf}). After the steady state was reached, a pre-designated amount of wood pellets (normally 20-40 g) containing known concentrations of sulfur (normally 0 or 4%) and a metal (normally 1000 ppm) was charged into the incinerator at a constant rate of about 1 g/s for combustion. At the conclusion of the experiment, the bed residue (wood ash and bed sorbent) and cyclone ash were collected and analyzed for their metal and sulfur concentrations.

Experimental parameters included metal type, concentrations of sulfur, incinerator temperature, and air flow rate. Metal species included lead, cadmium, and chromium.

Chemical Analysis

Metal concentrations in the combusted bed residue and cyclone fly ash were measured employing an HF-modified EPA Method 3050 [10]. The operation involved the microwave digestion (CEM Model MDS-81D) of metals from a solid sample using nitric acid, hydrochloric acid, hydrogen peroxide, and modified with hydroflouric acid. The resultant metal solution was then analyzed for its metal concentration using an atomic absorption spectrometer (AAS). A Perkin-Elmer Model 2100 Atomic Absorption Spectrometer was used in this study. Toxicity Characteristics Leaching Procedure (TCLP) tests were also performed to determine the leachability characteristics of the deposited metals associated with the operations.

Data Analysis

In addition to sulfur capture, the following two metal capture properties were calculated:

Specific Capture Capacity (ϕ) - This is defined to be the amount of metal captured by a unit mass of sorbent. It was calculated as:

$$\phi_b = C_b - C_o \quad (\text{or}\ \phi_f = C_f - C_o) \quad (1)$$

Percent Capture (ψ) - This is defined to be the percent of metal captured by fluidized bed sorbents relative to the total amount of metal charged. It was calculated as:

$$\psi_b = \frac{\phi_b \times W_b}{C_w \times W_w} \times 100\% \quad (2)$$

$$\psi_f = \frac{\phi_f \times W_f}{C_w \times W_w} \times 100\% \quad (3)$$

and

$$\psi_T = \psi_b + \psi_f \quad (4)$$

RESULTS AND DISCUSSION

Simulation Results

Tables 1 and 2 summarize the simulation results for lead and cadmium, respectively. Note that, in the simulation, $C_{12}H_{22}O_{11}$ was used as the fuel, the amount of metal in the fuel was 1%, the amount of sulfur was either 0 or 4%, and the percent excess air was 200%. The results shown in Table 1 for lead strongly indicate that, without the existence of sulfur, lead will react with silica to form $Pb_2SiO_4(s)$; it, however, will not react with calcium oxide. With the co-existence of sulfur and silica, most of the lead will react with sulfur to form $PbSO_4(s)$. For cadmium, the results shown in Table 2 indicate that, without the existence of sulfur, cadmium is thermodynamically preferred to react with silica and alumina to from $CdSiO_4(s)$ and $CdAl_2O_4(s)$, respectively; it, however, will not react with calcium oxide. Similar to that occurs to lead, with the co-existence of sulfur and silica, most of the cadmium will react with sulfur to form $CdSO_4(s)$. The existence of sulfur, however, will not affect the reaction between cadmium and alumina in forming $CdAl_2O_4(s)$. More detailed description of equilibrium speciation showing the effect of sulfur is displayed in Figs. 2 through 5.

Experimental Results

Selected experimental results indicating the effect of temperature on simultaneous metal and sulfur capture by lime are shown in Figs. 6 through 8 for lead, chromium and cadmium, respectively. The following experimental parameters apply to these results: wood pellets amount - 20 g, metal type - nitrate, metal concentration - 1000 ppm, lime amount - 300 g, lime size - 0.5 mm, and air flow rate - 30 cm/s.

The results shown in Fig. 6 for simultaneous lead and sulfur capture indicate that the capture process is very efficient mostly above 80%. The existence of sulfur increases the efficiency of lead capture at a temperature below 750°C and decreases the efficiency above this temperature. In addition, the capture of sulfur by lime is seen to increase with the temperature up to about 85% at 900°C. Since lime is not expected to chemically react with lead according to equilibrium calculations, the mechanism of lead capture by lime is likely to be through "melt capture" as reported by Linak and Wendt [11]. The decrease in the efficiency of lead capture by lime with the existence of sulfur at 900°C is probably due to the decrease in "capture sites", i.e., many of the available metal capture sites are occupied by $CaSO_4$ (the reaction product of CaO and SO_2). The increase of lead capture by lime with the existence of sulfur at a temperature below 750°C apparently indicates that lime is capable of scrubbing $PbSO_4(s)$ efficiently.

For chromium capture by lime, the results shown in Fig. 7 indicate that the capture pattern is very similar to that of lead capture by lime. Since chromium is not expected to chemically reacted with lime either, the same discussion in the previous paragraph regarding lead capture by lime may also be applicable to chromium. The same discussion, however, does not apply to cadmium capture by lime as shown in Fig. 8. These results indicate that lime is not an effective sorbent for capturing cadmium nitrate. This observation is consistent with our previous reports [3].

It is worth pointing out that good fluidization is essential in the capture process. The capture efficiency was observed to be much lower when the bed was operated near the packed bed conditions.

CONCLUSIONS

The effects of sulfur on metal capture by lime during fluidized bed combustion were experimentally and theoretically investigated. The results indicated that the existence of sulfur would affect the metal capture efficiency. The level and the nature of the effect, however, depends on the metal species and the combustion temperature. Lime was observed to be capable of effectively capturing sulfur, chlorine and selective metals simultaneously.

ACKNOWLEDGEMENTS

The authors are grateful for the financial support of this study from USEPA through Gulf Coast Hazardous Substance Research Center (Grant No. 094LUB0399) and from USDOE through Pittsburgh Energy Technology Center (Grant No. DE-FG22-94PC94221).

NOTATION

C_b	metal concentration in bed sorbent, ppm
C_f	metal concentration in cyclone sorbent, ppm
C_o	metal concentration in original sorbent, ppm
C_w	metal concentration in wood pellets, ppm
d_p	particle diameter, mm
T	incinerator temperature, °C
U	air superficial velocity, cm/s
W_b	weight of collected bed sorbent, g
W_f	weight of collected cyclone sorbent, g
W_w	weight of wood pellets, g
ϕ_b	specific capture capacity of bed sorbent, mg/g
ϕ_f	specific capture capacity of cyclone sorbent, mg/g
ψ_b	percent capture by bed sorbent, %
ψ_f	percent capture by cyclone sorbent, %
ψ_T	total percent capture by sorbent, %

LITERATURE CITED

1. Oppelt, E. T., "Incineration of Hazardous Waste - A Critical Review," JAPCA, **37**, No. 5, 558-586 (1987).

2. Oberacker, D. A., J. J. Cudahy, and M. K. Richards, "Remediation (Clean-Up) of Contaminated Uncontrolled Superfund Dumpsites by Incineration and Other Popular Technologies," U.S.EPA, THERMAL1.DOC, 09/04/90.

3. Ho, T. C., C. H. Chen, J. R. Hopper and D. Oberacker, "Metal Capture During Fluidized Bed Incineration of Wastes Contaminated with Lead Chloride," Combustion Science and Technology, **85**, 101 (1992).

4. Ho, T. C., H. T. Lee, H. W. Chu and J. R. Hopper, "Metal Capture by Sorbents during Fluidized Bed Combustion," Fuels Processing Technology, **39**, 373 (1994).

5. Ho, T. C., R. Ramanarayan, J. R. Hopper, W. D. Bostick, and D. P. Hoffman, "Lead and Cadmium Capture by Various Sorbents during Fluidized Bed Combustion/Incineration," Proceedings of Fluidization VIII, p. 899, held in Tours, France May 14-19, 1995.

6. Lee, S. H. D. and I. Johnson, "Removal of Gaseous Alkali Metal Compounds from Hot Flue Gas by Particulate Sorbents," J. of Eng. for Power, **102**, 397 (1980).

7. Punjak, W. A., M. Uberoi and F. Shadman, "High-Temperature Adsorption of Alkali Vapors on Solid Sorbents," AIChE J., **35**, 1186 (1989).

8. Uberol, M. and F. Shadman, "Sorbents for Removal of Lead Compounds from Hot Flue Gases," AIChE J., **36**, 307 (1990).

9. Uberol, M. and F. Shadman, "High-Temperature Removal of Cadmium Compounds Using Solid Sorbents," Environmental Science & Technology, **25**, 1285 (1991).

10. Gao, D. and G. D. Silcox, "The Effect of Treatment Temperature on Metal Recovery from a Porous Silica Sorbent by EPA Method 3050 and by an HF Based Method," Air and Waste, **83**, 1004 (1993).

11. Linak, William P. and Jost O. L. Wendt, "Toxic Metal Emissions from Incineration: Mechanisms and Control," Prog. Energy Combust. Sci, **19**, 145 (1993).

Table 1. Equilibrium Simulation Results for Lead with or without Sulfur

Sorbent Constituent	Metal	With or Without Sulfur	Sulfur-Metal-Sorbent Compound	
SiO_2	Pb	Without S	$Pb_2SiO_4(s)$	< 900°C
			$PbO(g)$	> 900°C
		With S	$PbSO_4(s)$	< 920°C
			$Pb_2SiO_4(s)$	< 950°C
			$PbO(g)$	> 950°C
CaO	Pb	Without S	$PbO(s)$	< 850°C
			$PbO(g)$	> 850°C
		With S	$CaSO_4(s)$	> 500°C
			$PbSO_4(s)$	< 730°C
			$PbO(s)$	< 820°C
			$PbO(g)$	> 820°C

Table 2. Equilibrium Simulation Results for Cadmium with or without Sulfur

Sorbent Constituent	Metal	With or Without Sulfur	Sulfur-Metal-Sorbent Compound	
SiO_2	Cd	Without S	$CdSiO_3(s)$	< 830°C
			$CdO(s)$	< 920°C
			$Cd(g)$	> 920°C
		With S	$CdSO_4(s)$	< 780°C
			$CdSiO_3(s)$	< 830°C
			$CdO(s)$	< 870°C
			$CdS(g)$	> 870°C
Al_2O_3	Cd	Without S	$CdAl_2O_4(s)$	< 920°C
			$Cd(g)$	> 920°C
		With S	$CdAl_2O_4(s)$	< 930°C
			$CdS(g)$	> 930°C
CaO	Cd	Without S	$CdO(s)$	< 920°C
			$Cd(g)$	> 920°C
		With S	$CaSO_4(s)$	> 500°C
			$CdSO_4(s)$	< 730°C
			$CdO(s)$	< 920°C
			$Cd(g)$	< 940°C
			$CdS(g)$	> 940°C

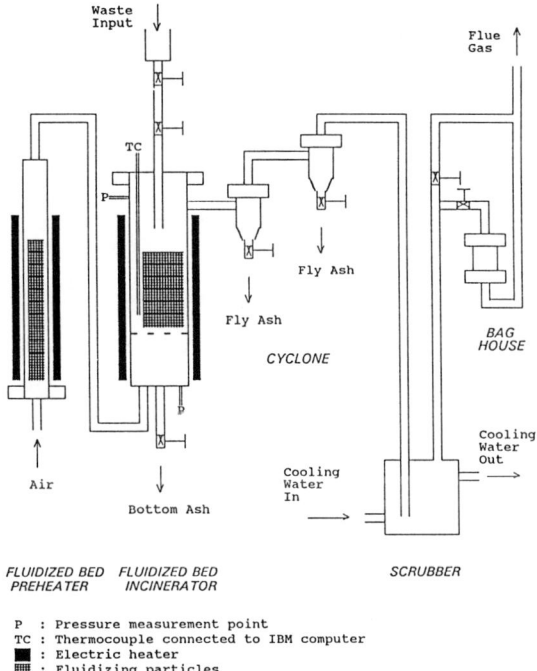

Fig. 1. Fluidized bed combustion system.

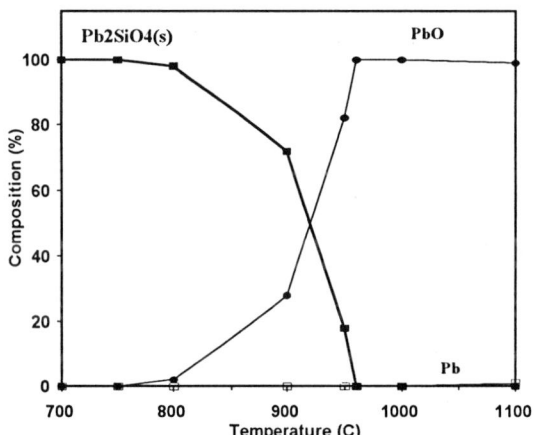

Fig. 2. Simulated lead speciation (System: Pb-SiO_2).

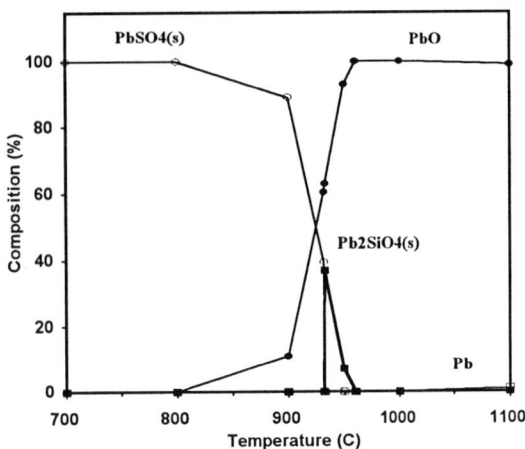

Fig. 3. Simulated lead speciation (System: Pb-S-SiO$_2$).

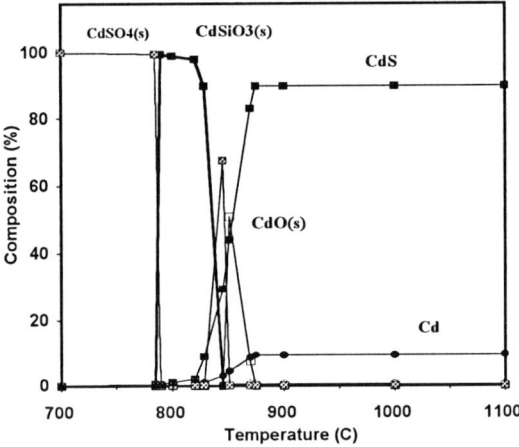

Fig. 4. Simulated cadmium speciation (System: Cd-S-SiO$_2$).

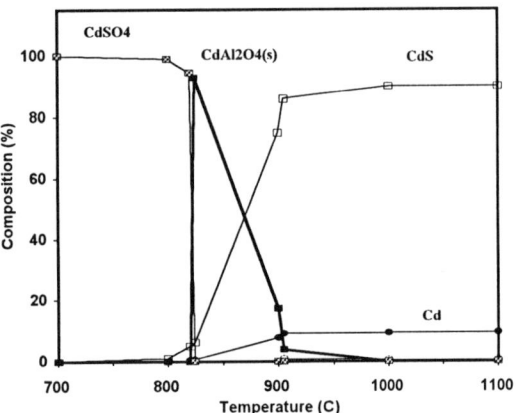

Fig. 5. Simulated cadmium speciation (System: Cd-S-Al$_2$O$_3$).

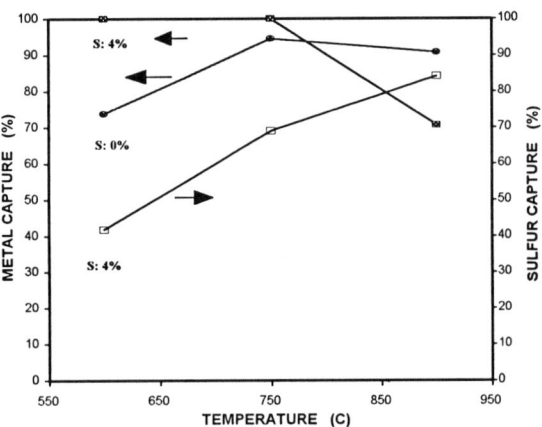

Fig. 6. Simultaneous lead and sulfur capture by lime.

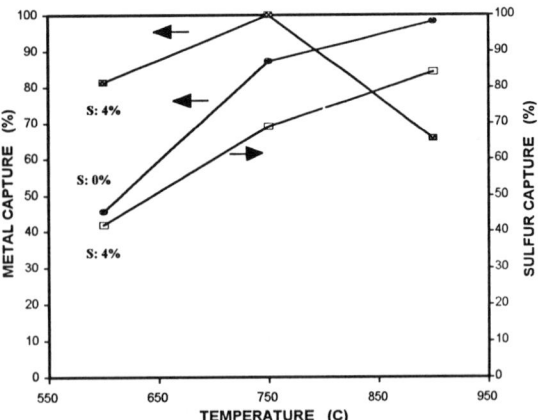

Fig. 7. Simultaneous chromium and sulfur capture by lime.

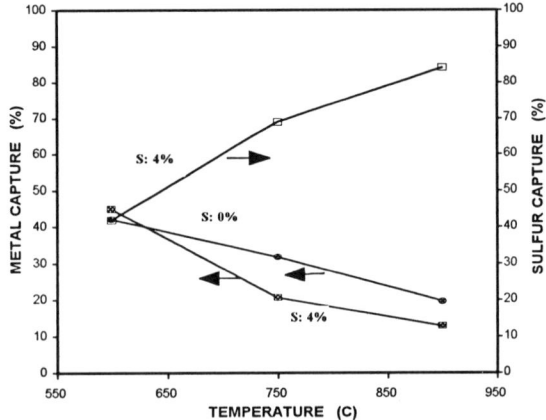

Fig. 8. Simultaneous cadmium and sulfur capture by lime.

Wall-to-Bed Heat Transfer in a Turbulent Fluidized Bed

Lii-ping Leu, Y. K. Hsia and C. C. Chen
Department of Chemical Engineering National Taiwan University, Taipei 106-17, Taiwan

The wall-to-bed heat transfer in turbulent fluidized beds was investigated in a 10.8 cm i.d. x 5.7 m high with expanded top fluidized bed. The Group A and Group B particles were used for experiments. The results showed that for Group A particles the heat transfer coefficient increased with the increase of superficial gas velocity. For Group B particles the heat transfer coefficient decreased with the increase of superficial gas velocity. The dependence of heat transfer coefficient on local voidage was also discussed.

INTRODUCTION

In the literature, heat transfer in high velocity fluidization is mainly focused on circulating fluidized beds, and research on heat transfer in turbulent fluidized beds is limited. Canada and McLaughlin [1] investigated the heat transfer of a tube bank to a fluid in a high pressure turbulent fluidized bed with 650 and 2600 μm particles. Staub [2] proposed a solids flow and heat transfer model and found satisfactory agreement with immersed tube bank heat transfer data. Ku et al. [3] presented an experimentally determined heat transfer coefficient in a turbulent fluidized bed using silica sand particles with mean diameter of 1200 and 3570 μm. Staub [4] conducted experiments on the flow behavior, bed expansion, and heat transfer coefficient to horizontal, in-bed tube banks in a three-dimensional square bed with 650 and 2600 μm glass particles and the same size silica sand. Hashimoto et al. [5] measured the heat transfer coefficient to the surface of vertical tubes located in the freeboard of a turbulent fluidized bed by using 48 μm and 2800 kg/m^3 bed material and proposed an empirical correlation. In this study, the heat transfer coefficient in turbulent fluidized beds was measured for Group A and Group B particles.

EXPERIMENTAL

The experiment apparatus is shown in Fig. 1. Experiments were conducted in a 10.8 cm i.d. x 5.7 m high three dimensional fluidized bed with a 25 cm i.d. x 1.5 m high expanded top section to reduce the gas velocity, hence the reduction of entrainments. According to Avidan and Yerushalmi [6] in this way solids entrainments falling back to the bed even at high velocities. The entrained particles from the expanded top section were collected by the cyclone and were recycled to the main bed simultaneously. The pressure taps were mounted flush with the wall of the column. A 3 mm i.d. copper tube were used as the pressure probe, one end of the probe was covered with a fine screen to prevent solid particles flowing out from the bed, and the other end of the probe was connected to a pressure transducer. A continuous pressure signal from the pressure transducer was amplified and sent it via an A/D converter to a personal computer for recording and for subsequent analysis. Particles used in experiments are shown in Table 1. A heat flux meter located at 40 cm above the distributor was used as a thermal probe to measure the heat transfer coefficient, the details of heat flux meter is shown elsewhere [7] and will not be repeated here.

RESULTS AND DISCUSSION

Transition velocities U_c and U_k

The transition velocities U_c and U_k were determined from the mean amplitude of pressure drop fluctuations by using a differential pressure transducer. Table 1 shows the result of U_c and U_k when the pressure tap spacing of

Table 1. Physical properties of particles used and experimental values of U_c and U_k.

solid particles	mean size(μm)	density(kg/m^3)	Geldart's classification	U_c(m/s)	U_k (m/s)
FCC	71	1800	A	0.38	0.45
sand	100	2635	B	0.55	-
sand	163	2635	B	0.8	-
sand	230	2635	B	1.0	-
sand	273	2635	B	1.35	-
sand	323	2635	B	1.48	-
sand	460	2635	B	1.98	-

a transducer is 10 cm. U_c and U_k were obtained for FCC Group A particles, but for 100 μm Group A particles only U_c was obtained. For Group B particles only U_c was obtained although a differential pressure transducer was used. These results were similar to the results obtained by Chehbouni et al. [8].

Heat transfer coefficient

For Group A particles, the heat transfer coefficient measured is shown in Fig. 2 as the function of superficial gas velocity. The heat transfer coefficient would increase as the superficial gas velocity increased. The heat transfer coefficient measured for turbulent fluidized beds was lower than the data obtained by Wunder [9] and Prins et al.[10] for Group A particles. And it was higher than the data for a circulating fluidized bed of same size FCC particles [7]. In the same figure, it was found that the heat transfer coefficient would increase as the particle size became smaller. The small particles had a large contact surface area per unit volume which shortened the transfer paths between particles and the heat transfer area, but this effect competed with the increase in voidage at same superficial gas velocity. The net effect would depend on which effect was stronger.

Fig. 3 shows the relationship between heat transfer coefficient and superficial gas velocity for Group B particles. The heat transfer coefficient measured decreased as the superficial gas velocity increased. The data here were higher than the results of Wunder [8]. The heat transfer coefficient would decrease as the particle size increased.

The effect of bed voidage on heat transfer coefficient is shown in Fig. 4 for Group A particles. In this study the heat transfer coefficient increased with the increase in bed voidage. Fig. 5 shows the relationship between heat transfer coefficient and bed voidage for Group B particles. The heat transfer coefficient decreased with an increase of bed voidage. The increase of bed voidage came with an increase in superficial gas velocity. The bed particles were carried upward vigorously by the gas, the chance for heat exchange became small, the heat transfer coefficient decreased.
Also the fact that the large particles had a low heat transfer coefficient is also shown in the same figure.

CONCLUSION

The heat transfer coefficient in turbulent fluidized beds was measured. For Group A particles, the heat transfer coefficient increased with an increase in superficial gas velocity; for Group B particles the heat transfer coefficient decreased with an increase in superficial gas velocity. For the Group A and Group B particles large size particles have smaller heat transfer coefficient.

NOTATION

h_w: heat transfer coefficient between wall and bed (W/m^2 °C)

U_c: superficial gas velocity at maximum amplitude of pressure fluctuations (m/s)

U_k: superficial gas velocity at which pressure fluctuations level off (m/s)

LITERATURE CITED

1. Canada, G. S. and M. H. McLaughlin, "Large Particle Fluidization and Heat Transfer at High Pressures", AIChE. Symp. Ser., no.176, vol.74, p. 27 (1978).

2. Staub, F. W., "Solids Circulation in Turbulent Heat Transfer to Immersed Tube Banks", J. Heat Trans., 101, 391 (1979).

3. Ku, A. C., M. Kuwata and F. W. Staub, "Heat Transfer to Horizontal Tube Banks in a Turbulent Fluidized Bed of Large Particles", AIChE. Symp. Ser., no. 208, vol.77, p. 359 (1981).

4. Staub, F. W., "Flow and Heat Transfer in Large Particle Turbulent Fluidized Beds", Proc. of Joint meeting of CIESC and AIChE, p.392, Peking (1982).

5. Hashimoto, O., S. Mori, S. Hiraoka, I. Yamada, T. Kojima and K. Tsuji, "Heat Transfer to the Surface of Vertical Tubes in the Freeboard of a Turbulent Fluidized Bed" Int. Chem. Eng., 30, 254 (1990).

6. Avidan, A. and J. Yerushalmi, "Solids Mixing in an Expanded Top Fluid Bed", AIChEJ., 31, 835 (1985).

7. Chen, C. C. and C. L. Chen, "Experimental Study of Bed-to-Wall Heat Transfer in a Circulating Fluidized Bed", Chem. Eng. Sci., 47, 1017 (1992).

8. Chehbouni, A., J. Chaouki, C. Guy and D. Klvana, "Characterization of the Flow Transition Between Bubbling and Turbulent Fluidization", Ind. Eng. Chem. Res., 33, 1889 (1994).

9. Wunder, R., Dissertation, TU, Munchen (1980). Cited by Molerus, O. and W. Mattmann, "Heat Transfer in Gas Fluidized Beds, part 2", Chem. Eng. Technol., 15, 240 (1992).

10. Prins W., G. J. Harmsen, P. de Jong and W. P. M. van Swaaij, "Heat Transfer from an Immersed Fixed

Silver Sphere to a Gas Fluidized Bed of Very Small Particles", in "Fluidization VI" (Grace, J. R., L. W. Shemilt and M. A. Bergougnou, ed.), p.677, Engineering Foundation, New York, New York(1989).

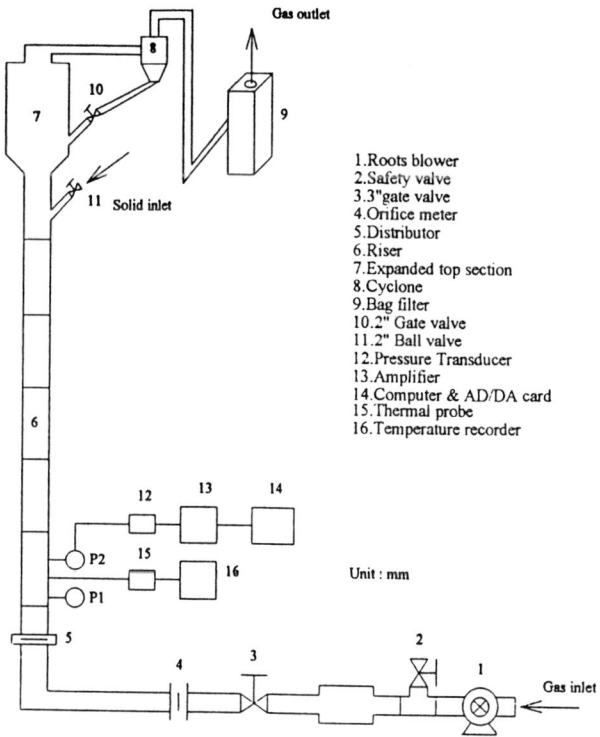

Fig. 1. Experimental Setup.

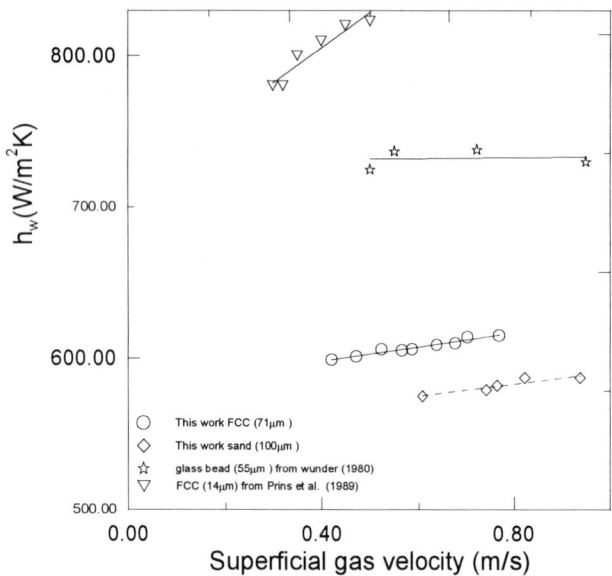

Fig. 2. Bed-to-wall heat transfer coefficient of Group A particles in turbulent fluidized beds.

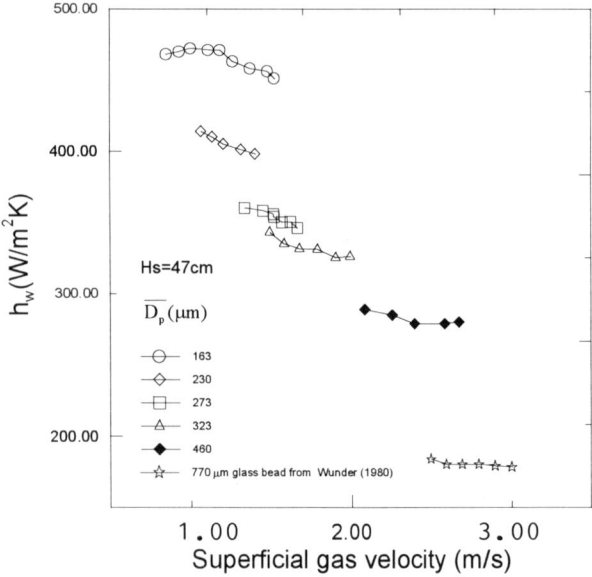

Fig. 3. Bed-to-wall heat transfer coefficient of Group B particles in turbulent fluidized beds.

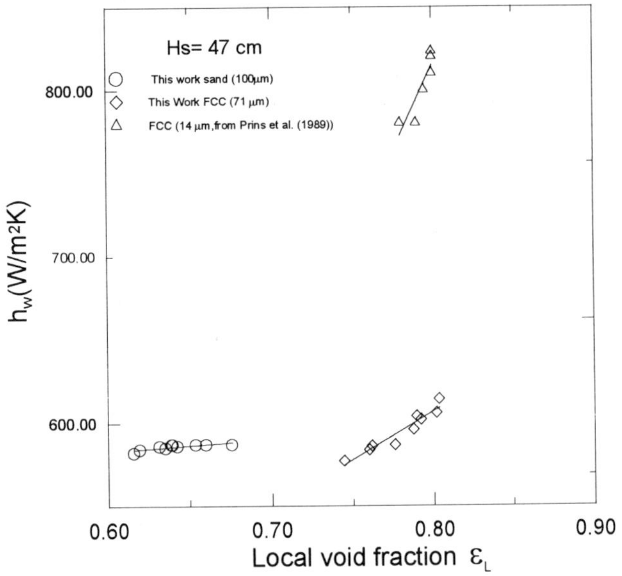

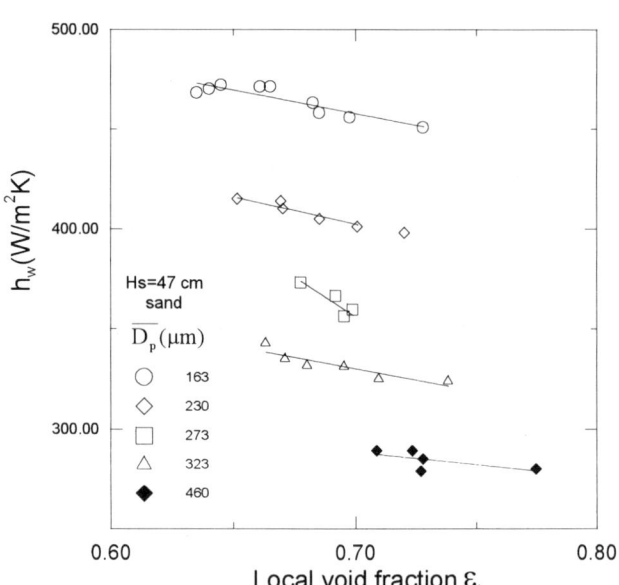

Fig. 4. Bed-to-wall heat transfer coefficient vs. local void fraction of Group A particles in turbulent fluidized beds.

Fig. 5. Bed-to-wall heat transfer coefficient vs. local void fraction of Group B particles in turbulent fluidized beds.

Experimental Study of a Zigzag Air Classifier

X. Loison, A. Sahnoun, P. Guigon and J.F. Large

Université de Technologie de Compiègne, Département Génie Chimique
Division Technologie des Poudres et Suspensions, B.P.20529. 60205 Compiègne Cedex, France

Air classifiers separate powders and granular materials into a fine and a coarse product in an air flow. The objective of this study is the characterization of a laboratory zigzag air classifier. The effects of several process parameters (feed rate, cut diameter) on the classification performance were studied for three solids with different properties (sand, sugar and salt).

INTRODUCTION

Air classification is defined as the process of sorting, arranging or allocating dry powders or particles by air flow, into two classes at least, to achieve a desired degree of uniformity. In powder technology, that means arranging particulate solids into classes in such a way that all the particles in a particular class exhibit similar behaviour properties, or that the class itself exhibits a given or a desirable behaviour pattern [1]. One can classify powders on the basis of several possible particles characteristics. The behaviour of particle collectives can be related to one or more properties of the individual entities in the collective [2]. Although many types of classification are possible, such as heavier (or more exactly : denser) particles from lighter ones and spherical particles from oblong or flat ones, the most common use of air classification in powder technology is in the separation of small particles from large ones.

The zigzag air classifier is a gravity classifier with a zigzag shaped channel. This channel contains a certain number of segments with a rectangular cross section fixed together at a given angle. In general, a gravitational flow of material and countercurrent air flow through this channel are the characteristics features of the zigzag classification principle. In this case, the coarse particles leave the classifier at the bottom outlet, whereas the fine ones move upwards with the air flow.

The separation performance and throughput capacity of a zigzag air classifier are determined by the particle behaviour at the individual stages and by the interaction between the stages that results from that single stage behaviour [3]. Many operating coupled parameters (feed rate, feed composition, air velocity, classifier geometry, particle size, particle density, particle shape...) influence the quality of the final product. The aim of the present paper is to discuss the influence of some of these parameters on the classification performance and quality.

EXPERIMENTAL SET UP

The pilot used in this study is a 120° zigzag air classifier made of stainless steel (Fig.1). The front section is constructed with a clear plastic and consists of twenty-seven rectangular segments with a fixed length of $5 \cdot 10^{-2}$ m. The width of these segments can be adjusted between 2 and $3 \cdot 10^{-2}$ m. The powder is fed into the classifier, at the 8th junction, by means of a vibrating feeder governed by a potentiometer which enables the operator to control the feed rate. The classification air, called primary air flow, is introduced at the bottom of the classifier. In order to transfer the fine fraction to a cyclone, a secondary air flow is introduced at the classifier top. The experiments were carried out with sand particles (2650 kg/m³), sugar (1581 kg/m³) and salt (2160 kg/m³) in order to have a wider range of solid characteristics. These three solids have a distribution size between 40 and 315 µm. In each segment, one can distinguish two kinds of particle flows : an upward stream of fine particles caused by the primary air flow and a downward stream of coarse material along the lower wall of the channel. All the particles of both streams are reclassified at every junction of the segments : they can be unaffected and continue their motion in the same direction or they can move backwards.

The upward mean air velocity in the classifier defines the cut size of the fed solid. The latter being chosen, the theoretical cut size velocity is considered to be equal to the terminal free fall velocity (Ut) which can be given by :

$$U_t = \sqrt{\frac{4(\rho_s - \rho_g) g d_p}{3 \rho_g C_d}}$$

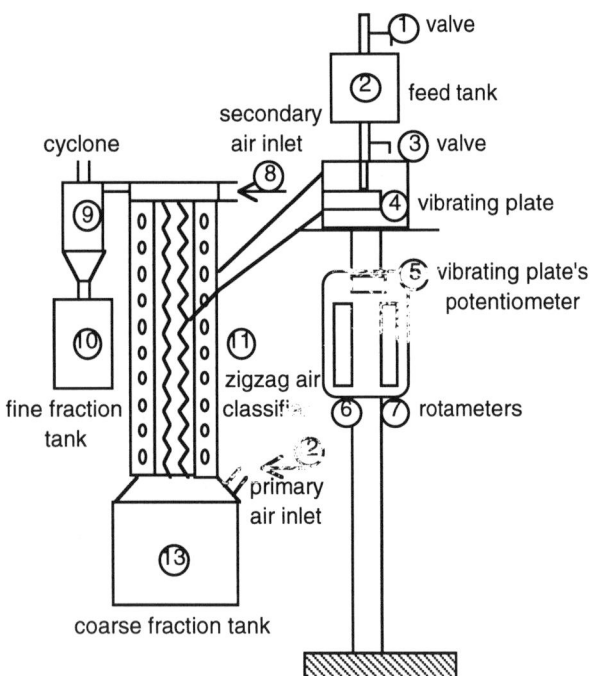

Fig. 1 : Schematic experimental set up

EXPERIMENTAL RESULTS

Evolution of the experimental cut size
For a fixed width of the classifier (2 10⁻² m) and a dilute particle flow (1.7 kg/s), it has been observed that the ratio between the experimental cut diameter (obtained by screening) and the theoretical one (obtained by the theoretical terminal free fall velocity) is almost constant for the three tested solids. This ratio is about 0.8. This value can be explained by the fact that, in the theoretical calculation of the terminal free fall velocity, we have considered that the particles were spherical and isolated.

For the same solid (sand), we have an increase of the ratio between the experimental and theoretical cut diameter when the width of the zigzag air classifier is increased (Fig. 2). This is probably due to the primary air velocity calculation corresponding to the desired cut size. It is based on the principle of terminal free fall behaviour which considers that the interactions between the particles and the walls are negligible. So, when the width of the zigzag classifier increases, the particle-wall friction becomes less important and consequently, the experimental cut diameter approaches the theoretical one.

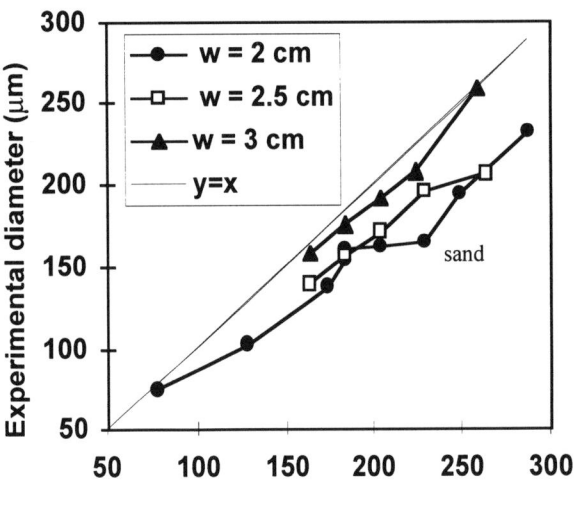

Fig. 2 : Comparison between the experimental and the theoretical cut diameter

Cut imperfection
The cut imperfection (I) leads to a characterisation of the quality of the desired fraction (fine or coarse). According to [4], it can be calculated by the following expression :

$$I = \frac{d_{84} - d_{16}}{2 d_{50}}$$

For a dilute feed (10g/min) and a classifier width of 2 cm, the cut imperfection seems to be in the same range (about 15 %) for the three solids tested and for several cut sizes (Fig.3). The role of the zigzag air classifier width on the cut imperfection is minor even if it seems smaller when the segment width decreases [5]. This could be explained by the fact that, when the width of the classifier is reduced, the friction forces become more important and consequently the risk of carrying away the fines with the coarse particles is reduced. Figure 4 shows for a fixed width of the zigzag classifier (2 cm), that the ratio between the primary air rate and the sand feed rate is very important regarding the cut imperfection. Two zones can be defined : in the first

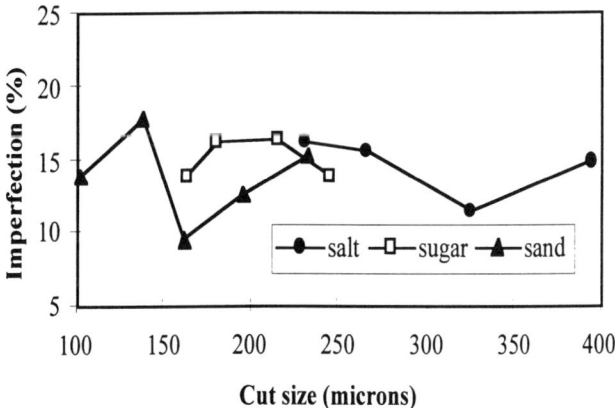

Fig. 3 : Influence of the cut size on imperfection

one, the imperfection value stays constant and in the second one, the imperfection increases regularly with the feed rate. It is clear that there is a critical value for the gas-solid ratio which influences the cut imperfection.

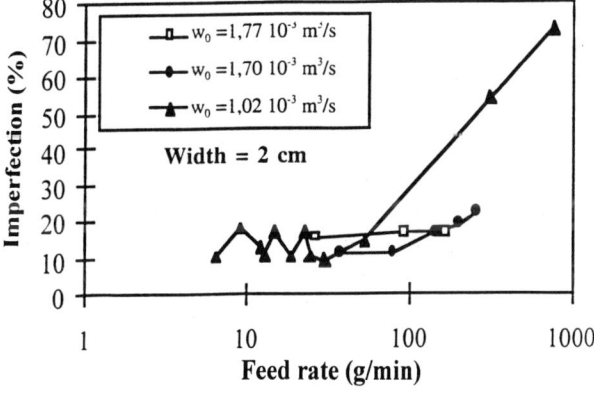

Fig. 4 : Influence of the solid rate on imperfection

Beyond this value the classification quality becomes worse. In order to operate in the appropriate conditions, it is necessary to determine this critical value for any application involving a different solid.

Residual of the fine in the coarse fraction

The residual of the fine in the coarse fraction allows to define the cut quality. It can be calculated by the following equation :

$$Rfg = \frac{Rmg \; Qg(d_{50})}{Qi(d_{50})}$$

According to figure 5, the increase of the desired cut size improves the quality of the coarse fraction and the residual in this fraction becomes smaller. Nevertheless, the classification quality is not the same for the three solids. The abrasive properties of these solids might have an influence since they increase the risk of carrying away the small particles with the coarse ones. Figure 6 shows that for a fixed sand rate (25 g/min), the quality of the coarse fraction gets better by reducing the width of the zigzag air classifier. This can be explained by friction forces which are more important as the width of the zigzag classifier decreases.

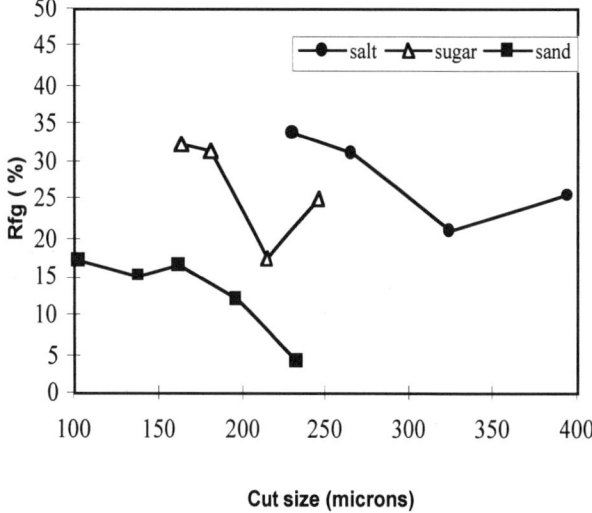

Fig. 5 : Influence of the solid nature on cut quality of coarse fraction

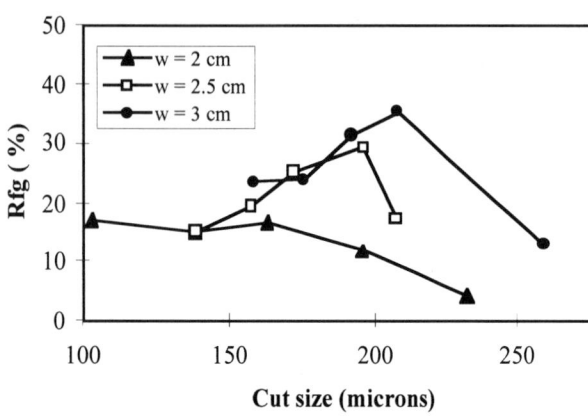

Fig. 6 : Influence of the zigzag width on cut quality of coarse fraction

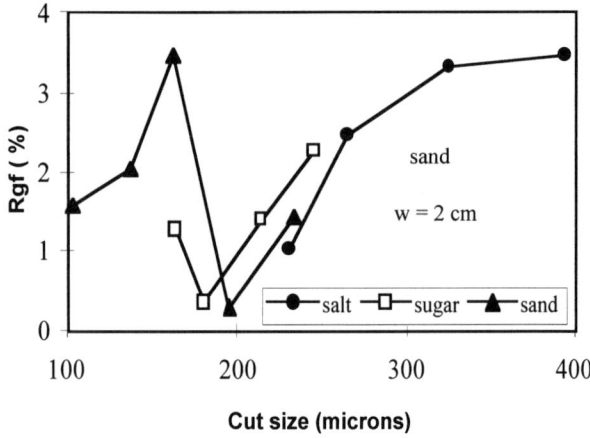

Fig. 7 : Evolution of the residual of coarse in the fine fraction

Residual of the coarse in the fine fraction

The residual of the coarse in the fine fraction enables us to characterise the cut quality of the fine fraction. It can be given by the following equation :

$$Rgf = \frac{Rmf\ (1 - Qf(d_{50}))}{1 - Qi(d_{50})}$$

According to figure 7, it seems that the residual of the coarse in the fine fraction varies in the same interval for the solids used in the zigzag air classifier. Compared to the residual of the fine in the coarse fraction (which was between 5 and 35 %) under the same operating conditions, the residual of the coarse in the fine fraction is small (less than 4 %) for all the solids tested. We could say that the use of zigzag air classifier leads to a good quality of the fine fraction. This can be explained by the fact that the fine or the air flow cannot easily carry away the coarse. Figure 8 shows again that the smallest width of the zigzag air classifier (2 cm) gives the best results. But, it has to be noted that even a width of 3 cm gives a better fine fraction quality than the quality of the coarse one.

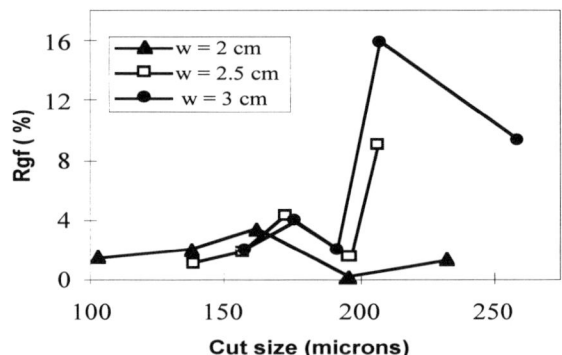

Fig. 8 : Influence of the zigzag width on the fine fraction quality

CONCLUSION

The aim of this paper was to describe some parameters influencing the zigzag air classification quality. Since the small particles cannot easily carry away the largest ones, the quality of the fine fraction is always better than the coarse fraction quality. This effect seemed valid for all used materials. The best classification quality was obtained with the minimal width used of the zigzag air classifier. For all tested solids, a threshold for the ratio between the primary air rate and the feed rate was observed. Exceeding this critical value leads to a decrease of the quality of both fractions.

List of symbols

C_d :	Drag coefficient	(-)
d_p :	Desired cut size	(m)
g :	Acceleration of gravity	(m s^{-2})
I :	Imperfection of cut	(%)
$Qf(x)$:	Cumulative weight fraction of the fine fraction at size x	(%)
$Qg(x)$:	Cumulative weight fraction of the coarse fraction at size x	(%)
$Qi(x)$:	Cumulative weight fraction of the feed at size x	(%)
Rfg :	Residual of the fine in the coarse fraction	(%)
Rgf :	Residual of the coarse size in the fine fraction	(%)
Rmf :	Weight fraction of the fine fraction	(%)
Rmg :	Weight fraction of the coarse fraction	(%)
Ut :	Terminal free falling velocity	(m s^{-1})
w :	Classifier width	(m)
$w0$:	Primary air flow rate	(m^3 h^{-1})
μ :	Air viscosity	(kg m^{-1} s^{-1})
ρ_g :	Air density	(kg m^{-3})
ρ_s :	Solid density	(kg m^{-3})

References

[1]   S. R. de Silva, Air classifiers : state of the art and future potential; Postec, Porsgrunn, Norway, 1994.

[2]   H. Rumpf, Staub, **27**(1), 1967, pp 3-13.

[3]   G.G. Rosenbrand, M.M.G. Senden and M. Tels, Proceedings of First World Congress Particle Technology, Part. IV, Nuremberg, Germany, 1986, pp 469-483.

[4]   Caractérisation des séparations granulométriques. Normalisation Française, XII-695, Novembre 1987, pp 484-524.

[5]   P. Guigon, X. Loison, A. Sahnoun, Etude d'un classificateur zigzag. PARTEC R95-2011. Internal Report.

The Elastic and Cohesive Properties of Particulate Beds

A.J. Matchett*, M. Aufauvre, J.M. Coulthard
University of Teesside, Middlesbrough, TS1 3BA, England

S. Alsop
Harwell Drying Restoration Services, Harwell, England

A non-destructive method of assessing elastic and cohesive properties of beds of powders has been developed based upon the measurement of resonance frequency of a powder bed. This has been applied to wet sands, coals and industrial chemicals. A Critical State model suggests that (a^2l/v) should relate linearly to $1/W$, where a is the resonant frequency (rad/s), l is sample height(m), v is specific volume (m^3/kg) and W is the top cap mass per unit area (kg/m^2). While data conform to this relationship, samples of differing size and shape give different lines. A shell model has been developed which assumes that all the strength of the material exists in a thin shell on the outside of the sample. This model implies that (a^2ld/v) is linearly related to l/W, where d is the sample equivalent diameter(m). Data for sand, coal and terephthalic acid(TA) each give a single straight line when plotted in this form for cylindrical samples of different sizes and cubic samples.

The data suggest that surface properties are important in determining the properties of beds of powders, rather than bulk properties. However, most methods of characterising beds of powders rely upon bulk properties.

INTRODUCTION AND CRITICAL STATE THEORY

Elastic properties of beds of powders are important in dynamic systems where the transmission of stress waves occur, and most materials must traverse the region of elastic deformation in order to reach sustainable plastic flow. This paper presents a non-destructive method of assessing elastic properties and their implications on the structure of beds of cohesive powders.

*Correspondence concerning this paper should be addressed to A.J. Matchett.

Norman-Gregory and Selig[1] developed a system of assessing the properties of beds of powders by subjecting them to vertical vibrations. Matchett[2] noted that their data could be analysed by a Critical State soil mechanics model. This was further developed by Alsop and Matchett[3].

The experimental equipment used by Alsop is shown in Figure 1. A sample of material was placed upon a vibrating table and surmounted by a top cap mass. Resonant frequency was found by measuring transmitted acceleration from the base to the top cap over a range of frequencies at constant acceleration. The frequency which gave the greatest amplification was the resonant frequency. For a given sample, this procedure was repeated at a number of top cap masses[3].

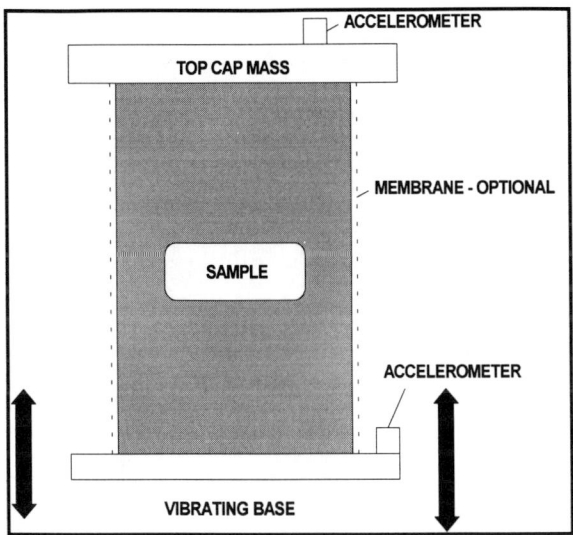

Figure 1 Alsop and Matchett Resonant Frequency Vibration Test[1][2]

The Critical State model gives the following relationship for resonant frequency a(radians/s):

$$a = \sqrt{\frac{3(1+2\alpha)p_s v_s}{Wkl_s}} \quad (1)$$

Where p_s is the static value of the effective compressive stress (Pa), v_s is the static value of specific volume (m³/kg), α is the ratio of radial strain to axial strain, W is top cap mass per unit area (kg/m²), k is the Critical State elastic swelling line gradient[4], l_s is the static value of bed height.
Where
$$v = v_k - k\ln(p') \quad (2)$$

p' is the effective compressive stress(Pa). u_c is the cohesive compressive stress which holds the sample together. It is assumed that u_c acts isotropically and obeys the principles of effective stress[4]. The static effective compressive stress is caused by the top cap mass and u_c and is given by:

$$p_s = \frac{(\sigma_1' + \sigma_2' + \sigma_3')}{3} = \frac{Wg}{3} + u_c \quad (3)$$

Thus

$$Q = \frac{a^2 l_s}{v_s} = \frac{3g}{k_1} + \frac{9u_c}{k_1}\left(\frac{1}{W}\right) \quad (4)$$

$$k_1 = \frac{3k}{(1+2\alpha)}$$

The small deformations inherent in elastic deformation have made it difficult to measure α directly, and hence equation 4 has been used to determine k_1 and u_c. Typical data are given for two sizes of cylindrical sample in Figure 2. While for each sample, the plot of Q versus 1/W give straight lines, the samples conform to different linear relationships, therefore the Critical State model does scale-up in a simple manner.

ELASTIC SHELL THEORY

Butensky and Hyman[5] proposed a model of wet agglomerates based upon a dry shell with a wet centre. If this model were valid for the materials used in these tests, it would suggest that the elastic properties of the sample would be confined in a thin shell around the surface of the sample - see Figure 3.
If the shell is considered as a thin elastic system in plane strain, with thickness δ (m) and modulus of elasticity E (Pa), then it can be shown that:

$$a^2 = \frac{4\delta E}{l_s W d} \quad (5)$$

or
$$Qd = \left(\frac{4\delta E}{v}\right)\frac{1}{W} \quad (6)$$

Where d is the sample diameter for cylindrical samples and the equivalent diameter for non-cylindrical samples = 4* cross sectional area/perimeter. Thus, equation 6 suggests that Qd should be plotted against 1/W. Data for wet sand are shown in Figure 4 for small and large cylinders and also cubic samples. Data for coal are given in Figure 5. These data fit the shell model. Details of the mould sizes are given in Tables 1
Figure 6 shows data for terephthalic acid (TA). Samples were made in conventional steel moulds and also in PTFE moulds in order to reduce wall friction - Table 1. This material also follows the shell model and the PTFE had no effect upon the results.

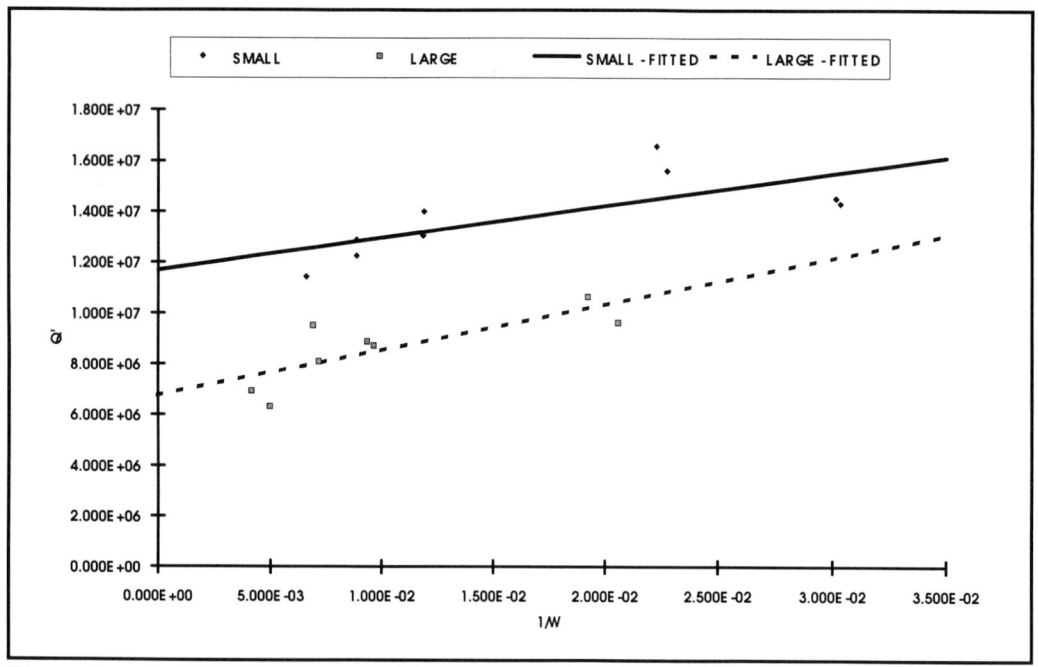

Figure 2 Plot of Q' versus 1/W for Wet Sand
Particle size <2mm : Bulk density 1500 kg/m^3 : moisture 3%
Q' is the value of Q with frequency expressed in Hz.

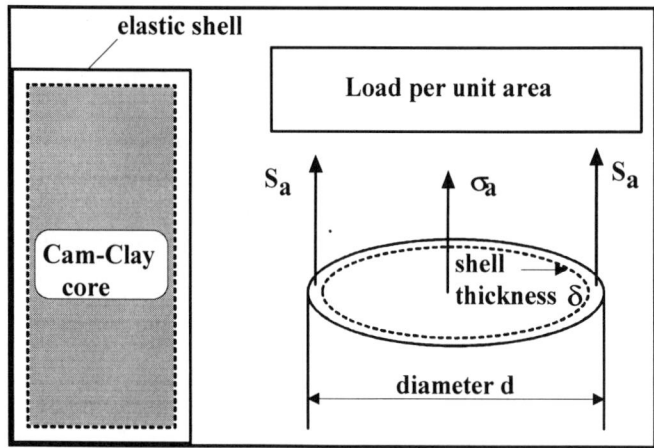

Figure 3 The Proposed Shell Structure

Table 1 Details of Moulds used in Experimental Programme

MOULD	SHAPE	SIZE mm	MATERIAL
Large	Cylindrical	100*50 dia.	Steel & PTFE
Small	Cylindrical	76*38 dia	Steel & PTFE
Cube	Cube	50*50*50	Steel

DISCUSSION

Within the limits of our experimental programme, the shell model describes the behaviour of a number of cohesive solid systems, in accordance with Butensky and Hyman's model.

There are several implications to these findings and scope for future work:

i) It has been demonstrated that the materials used have definite structures

ii) The controlling properties of these materials reside in a thin shell at the surface of the material - not in the bulk of the material.

iii) Most materials testing procedures measure bulk properties - if the shell model is applicable, they could be measuring the wrong parameters

iv) Further work needs to be done on the effects of sample size and shape - as size increases, the surface effects should become less noticeable

v) Wet materials have been used, where cohesive effects have been due to liquid bridge bonding. It would be interesting to see if other types of materials were subject to the same rules

vi) Only elastic properties have been studied. Would a shell model apply to plastic deformation and failure?

CONCLUSIONS

A shell model for bed structure has been tested on a range of materials, using a non-destructive method of assessing elastic properties from the resonant frequency of a bed of powder surmounted by a top cap mass. Materials tested include sands, coals and TA.
All these materials show that elastic properties are contained within a thin shell at the surface of the sample, thus indicating a definite sample structure. The implications of these findings have been considered.

REFERENCES

1. **Norman-Gregory G.M, T.E.Selig**, J.Geotech.Eng., 1987, **115**, 289
2. **Matchett A.J.**, Powder Technology, 1992, **70**, 63
3. **Alsop S.J., A.J. Matchett**, Shock and Vibration J., 1995, **2**, 383
4. **Schofield A., C.P.Wroth**, 'Critical State Soil Mechanics', McGraw-Hill, Reading, UK, 1969
5. **Butensky M., D.Hyman**, Ind.Eng.Chem.Fundam., 1971, **10**, 212

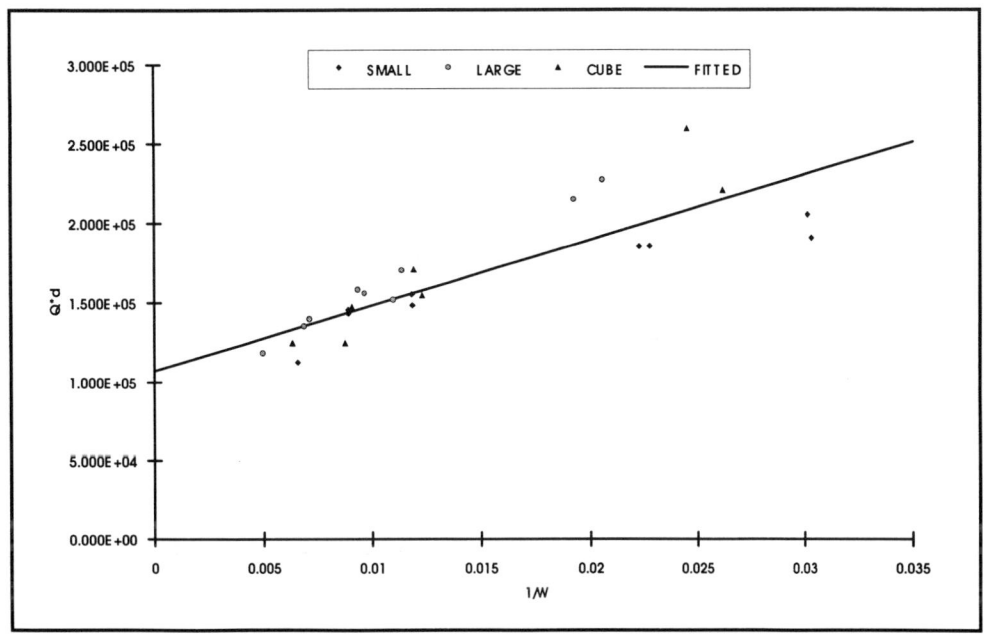

Figure 4 Plot of Q'd versus 1/W for Wet Sand
Particle size<2mm : Bulk density 1600 kg/m^3 : Moisture 6%

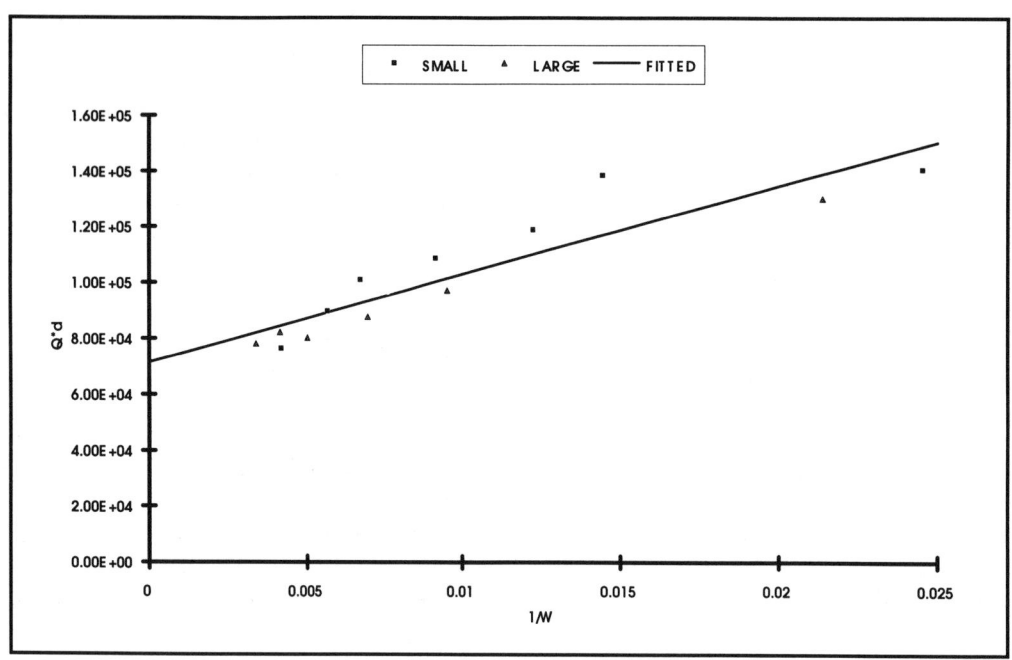

Figure 5 Q'd versus 1/W for Pitson McClure Coal
Particle size <1mm : Bulk density 1000 kg/m^3 : Moisture 5%

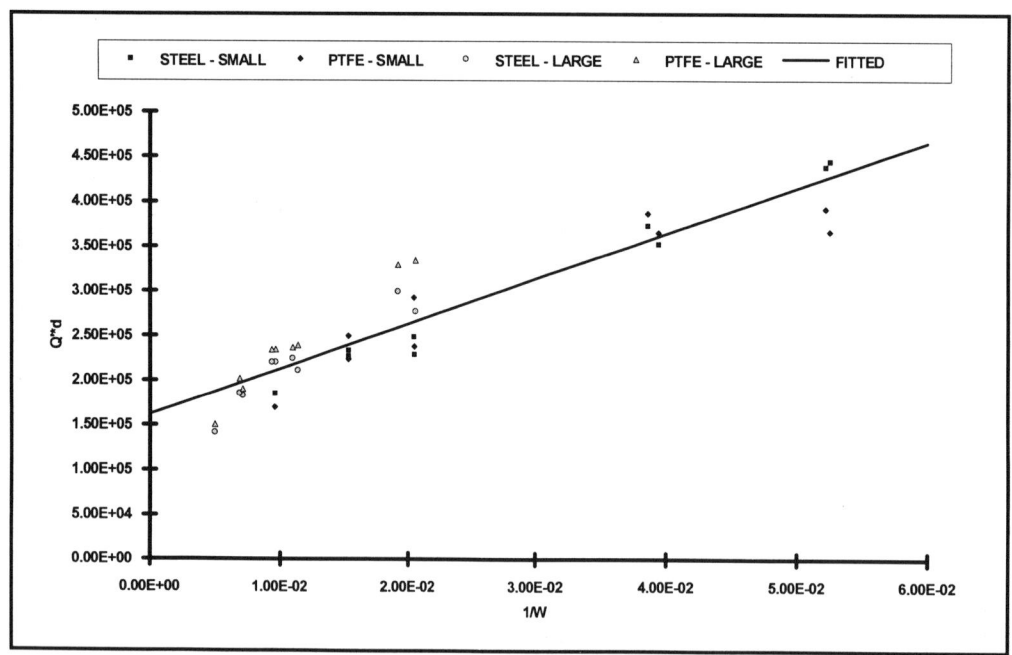

Figure 6 Q'd versus 1/W for TA
Particle size<100mm : Bulk density 1000 kg/m^3 : Moisture 9%

Axial Solids Distribution and Bottom Bed Dynamics for Circulating Fluidized Bed Combustor Application

Robert C. Zijerveld[1], Alain Koniuta[2], Filip Johnsson[3], Antonio Marzocchella[4]

Jaap C. Schouten[1] and Cor M. van den Bleek[1]

[1]Delft University of Technology, Chemical Reactor Engineering Section, Delft, The Netherlands
[2]Cercha, Rue Aimé Dubost, BP 19, 62670 Mazingarbe, France
[3]Chalmers University of Technology, School of Mechanical Enigneering, Göteborg, Sweden
[4]Universita di Napoli "Federico II", Dipartimento di Ingegneria Chimica, Naples, Italy

In this study simple models are used to describe both the axial solids distribution of Circulating Fluidized Bed (CFB) risers and the dynamic characteristics of the bubbles and "slugs" present in the bottom bed. Model results are compared with experimental data obtained in CFB risers of different size, all operated with 0.30 mm silica sand (i.e. Group B solids) at gas velocities typical for combustors ($U_g = 2 \div 4$ m/s). The wide risers ($D_r > 0.4$ m) have flow conditions representative for combustors, while the narrow risers have not. The bottom bed solids concentration is well described and the calculated bubble dynamics are in agreement with Kolmogorov entropy and frequency obtained from pressure fluctuations measured in the bottom bed of tile larger CFBs. The comparison of experiments and models indicates that the major part of the gas (>95 %) passes the bed other than as visible bubble flow. The decay constant a for the splash zone above the bottom bed increases significantly when riser size is increased. The decay constant a in narrow risers shows a strong dependency on riser solids holdup (and thus on solids mass flux) at constant superficial gas velocity.

Circulating Fluidized Bed (CFB) combustors are typically operated with Group B solids at superficial gas velocities which exceed the terminal velocity of an averaged sized bed particle but are not much higher than the terminal velocity of the largest size fraction of the particles. At these conditions there is considerable elutriation of solids, although solids mass fluxes are typically below 40 kg/m²/s and much lower than common in FCC riser applications. Generally, CFB combustors are operated with a bottom bed with a height typically less than 0.5 m.

The dynamics of the bottom bed in wide risers ($D_r > 0.4$ m) are of the exploding bubble type. Bubbles are large, compared to bubbles in the bubbling bed regime at low superficial gas velocities and explode vigorously at the bed surface, throwing solids into the splash zone. Since the bubbles are limited in size by the height of the bottom bed they are smaller than the diameter (or width) of the facility [1], i.e. slugging does not occur. A large amount of gas by-passes the bed, because when an exploding bubble has grown to its final size, it may reach from the gas distributor to the bed surface and a significant gas flow through the bubbles occurs. Hence, the bottom bed voidage is low with a value of about 0.6. Exploding bubbles do not have a well-defined shape as the bubbles in the bubbling bed at low superficial gas velocities, but can be seen as large voids of gas.

The dynamics of the bottom bed in narrow risers ($D_r < 0.4$ m) operated with a bottom bed at relatively high gas velocities (i.e. circulating conditions) are of the "slugging" type: the size of the voids is limited by the size of the cross-section of the facility. A similar phenomenon is well known for non-circulating conditions, for which wall limitations causes a bed to develop slugs, while the wide bed is still bubbling for similar bed solids holdup and superficial gas velocity. For circulating conditions in risers smaller than 0.4 m ID, Arena et al. [2] observed typical "plug-slug" structures in the presence of a bottom bed. These "plug-slug" structures can be considered as a result of wall-limitations and were found to have a high rise velocity (> 5 m/s). The gas "slugs" push the solids "plugs" with high velocity through the riser. In the transition zone the plugs loose solids to the wall zone, break up and finally the "plug-slugs" cannot be observed anymore in the transport or dilute zone. The dynamics of the bottom bed in narrow risers is significantly different from the dynamics of the dense bottom bed in wide risers, as shown by Zijcrvcld et al. [3] and thc bottom bcd voidagc is about 0.8. These flow conditions in the narrow risers are not representative for combustors.

Both in wide and narrow risers a splash or transition zone characterized by strong solids backmixing is present above the bottom bed. The solids concentration in this zone may be described with an exponential decay (e.g. Kunii and Levenspiel, [4]; Johnsson and Leckner [5]). Although the axial solids profile in both narrow and wide risers may be characterized in a similar way (i.e. by a bottom bed and an exponential decay), they are different from a *dynamic* point of view [3], so phenomena in the transition or splash zone are not necessarily similar. Above the splash zone a transport zone may be present, which is characterized by solids backmixing mainly at the riser wall. This zone is included in the model of Johnsson and Leckner [5]. The solids concentration may be described with an exponential decay, with a lower value of the decay constant than in the splash zone. Kunii and Levenspiel [4] model this zone with a constant solids concentration in case of low superficial gas velocities ("bubbling or turbulent fluidized bed") or this zone lacks in case of high superficial gas velocities ("fast fluidized bed"). This difference between the models of Kunii and Levenspiel [4] and Johnsson and Leckner [5] causes that the decay constant for the transition or splash zone in CFBs may have different values, depending on the model used to estimate it, although in both models this decay constant is indicated with the same symbol *a*.

The observation that the bottom bed in wide combustors is of a bubbling type encourages the use of a conventional bubbling bed model [6,7] extended with a description of the splash and transport zone [5] to model the hydrodynamics of CFB combustors. Here the model of Werther and Wein [7] is included for the bottom bed modelling, since they showed that void sizes and velocities found experimentally under conditions (U_g and Group B solids) typical for combustors, were in agreement with calculated bubble diameters and rise velocities.

The present work investigates whether it is possible to use existing models [e.g. 4,5,7] to predict the axial solids distribution of CFBs of different size operated at typical combustor gas velocities with the same Group B solids. Furthermore, it is investigated whether the bubble characteristics of the bottom bed predicted by the model are consistent with dynamical measurements carried out in the larger CFBs.

EXPERIMENTS

Experimental facilities

Table 1 shows the main characteristics of the five CFB facilities used. The 12 MW_{th} boiler was operated at 850 °C and the other four facilities were operated under ambient conditions. The 1.2x0.8 m facility is equipped with water manometers and the other facilities are equipped with pressure transducers along the riser height to obtain the axial pressure profile.

Experimental methods

The average axial voidage profile is calculated from the axial pressure profile neglecting friction effects and acceleration of the gas-solids suspension. From voidage and pressure profiles the bottom bed can be detected and its voidage determined. For all risers the solids holdup is directly obtained from the pressure profile; the pressure drop between the lowest pressure tap in the riser and the distributor is estimated by means of an extrapolation of the pressure profile assuming constant voidage in the bottom bed.

Measurements of pressure fluctuations were carried out at the wall in the bottom zone of two of the risers: at 0.35 m in the centre of the wall of the 12 MW_{th} boiler and at 0.28 m in the centre of the 1.2 m wall of the 1.2x0.8 m riser. The fluctuations of the pressure with respect to its local average were measured with piezoelectric transducers (Kistler, type 7261).

Average and peak frequency of the pressure fluctuations are determined by means of spectral analysis. The cycle frequency is calculated as the number of times that the pressure signal crosses its average value per time unit divided by two. Kolmogorov entropy is estimated by means of non-linear analysis. It is a measure for the predictability and is large for very irregular dynamics (*e.g.* turbulence) and small for more regular dynamics (*e.g.* slugging). The method used to estimate the Kolmogorov entropy from measured time series is described elsewhere [8].

Table 1 Characteristics of CFB facilities

Facility	Location	Distributor	Cross-section Size [m]	Shape	Height[m]
12 MW_{th} boiler	Chalmers	144 bubble caps	1.47x1.42	Rectangular	13.5
1.2x0.8 m CFB	Cerchar	54 bubble caps	1.20x0.80	Rectangular	9.0
0.70 x 0.12 m CFB	Chalmers	6000 holes	0.70x0.12	Rectangular	8.5
0.12 m ID CFB	Naples	various	0.12 ID	Circular	5.8
0.083 m ID CFB	Delft	porous plate	0.083 ID	Circular	4.0

Operating conditions

The silica sand used is characterized by d_p=0.30 mm, U_t=2.5 m/s (25 °C) and U_t=1.9 m/s (850 °C). Each run is characterized by superficial gas velocity (U_g) and riser solids holdup, the latter quantified by riser pressure drop (ΔP_{riser}). Superficial gas velocity ranges from below to beyond the terminal velocity and a bottom bed could always be identified.

MODEL

Table 2 shows the model equations used in this work. The bottom bed is modelled with equations (1) to (9), according to the model of Werther and Wein [7]. This part of the model calculates the bottom bed solids concentration $c_v(Z)$ and the bubble diameter $d_v(Z)$ at height Z above the gas distributor.

Equation (11) assumes the Kolmogorov entropy to be linearly proportional to the total number of bubbles that erupt at the bed surface per unit of time, $N_{bubble}(H_x)$, cf.[9]. Equation (12) assumes the bubbles to be spherical and equally sized. This part of the model takes the bottom bed dynamics into account.

The part of the model which calculates $c_v(Z)$ and the bubble diameter $d_v(Z)$ at height Z above the gas distributor is solved simultaneously with the part which calculates the bottom bed dynamics (N_{bubble} and K_{ML}). The proportionality constant ξ (the bubble impact factor [9]) in equation (11b) is systematically fitted with one value for all operating conditions and facilities.

The riser solids distribution above the bed is modelled according to i) Johnsson and Leckner [5], assuming a decay constant a for the splash zone and a decay constant K for the transport zone (equations (13) and (14)) and ii) Kunii and Levenspiel [4], assuming one decay constant a for the entire riser above the bottom bed (equation (15)). The riser pressure drop ΔP_{riser} can be obtained from an integration of equations (1) and (13) or (1) and (15).

RESULTS AND DISCUSSION

The upper part of Table 3 shows experimental and calculated c_{vx} for some characteristic runs carried out in the 1.2x0.8 m riser. Calculations are carried out with the φ and γ as suggested by Werther and Wein [7]. Clearly can be seen that the model with the values of φ and γ given by equations (5) and (9) is not able to describe the experimental c_{vx}. Similarly c_{vx} for the 12 MW_{th} boiler is not accurately described, which is not reported here. These discrepancies may exist, since the values for these parameters used were obtained for specific solids type and operating conditions. Therefore in this work two adjustments have been made. i) The value of the fraction of total gas flow in visible bubble flow, φ, has been lowered considerably. One value of φ is estimated in a systematic way for all operating conditions (U_g and H_x) and facilities. ii) The value of the

Table 2 Model equations

$$c_{v_x} = (1-\varepsilon_b)c_{v_d} \quad Z<H_x \quad (1)$$

$$\frac{c_{v_d}}{c_{v_{mf}}} = 1 - 0.14 N_{Re_p}^{0.4} N_{Ar}^{-0.13} \quad (2)$$

$$\varepsilon_b = \frac{\varphi(u_g - u_{mf})}{u_b} \quad (3)$$

$$u_b = \varphi(u_g - u_{mf}) + 0.71\vartheta\sqrt{g d_v} \quad (4)$$

$$\varphi = 1.45 N_{Ar}^{-0.18} \quad 10^2 < N_{Ar} < 10^4 \quad (5)$$

$$\vartheta = \begin{cases} 0.63 & d_t<0.1m \\ 2.0\sqrt{d_t} & 0.1m \le d_t \le 1.0m \\ 2.0 & d_t>1.0m \end{cases} \quad (6)$$

$$\frac{d(d_v)}{dZ} = \left(\frac{2\varepsilon_b}{9\pi}\right)^{\frac{1}{3}} \quad (7)$$

$$d_{v0} = \gamma \frac{(u_g - u_{mf})^{0.4}}{g^{0.2}} \quad (8)$$

$$\gamma = 1.3\left(\frac{A_{bed}}{N_{nozzle}}\right)^{0.4} \quad (9)$$

$$u_{slug} = \varphi(u_g - u_{mf}) + 0.35\sqrt{g d_t} \quad (10)$$

$$K_{ML} \propto \frac{\#\ bubble\ eruptions}{s} = N_{bubble} \quad (11a)$$

$$K_{ML} = \xi\ N_{bubble} \quad (11b)$$

$$N_{bubble}(H_x) = \frac{\varphi(u_g - u_{mf})A_{bed}}{\frac{\pi}{6}d_v^3(H_x)} \quad (12)$$

$$c_v = (c_{v_x} - c_{v_{2,x}})e^{-a(Z-H_x)} + c_{v_{exit}}e^{K(H_{exit}-Z)} \quad H_x<Z<H_{exit} \quad (13)$$

$$c_{v_{2,x}} = c_{v_{exit}}e^{K(H_{exit}-H_x)} \quad (14)$$

$$c_v = (c_{v_x} - c_v^*)e^{-a(Z-H_x)} + c_v^* \quad H_x<Z<H_{exit} \quad (15)$$

Table 3 Predicted and experimental bottom bed characteristics

			Model					Exper.
	U_g	ΔP_{dist}	φ	γ	d_{v0}	c_{vd}	$c_{vx}(H_x)$	c_{vx}
This work	2.0	2.6	0.344	0.259	0.212	0.43	0.34	0.45
param. W&W [5]	3.2	4.0	0.344	0.259	0.258	0.41	0.29	0.44
1.2x0.8 m CFB	4.0	5.3	0.344	0.259	0.283	0.39	0.26	0.42
This work	2.0	2.6	0.0175	0.325	0.266	0.43	0.42	0.45
param. W&W [5] adapted	3.2	4.0	0.0175	0.325	0.324	0.41	0.40	0.44
1.2x0.8 m CFB	4.0	5.3	0.0175	0.325	0.355	0.39	0.38	0.42
This work	2.0	1.3	0.0175	0.386	0.343	0.48	0.47	0.42
	3.2	2.4	0.0175	0.386	0.417	0.46	0.45	0.40
12 MW$_{th}$ boiler	4.0	3.9	0.0175	0.386	0.457	0.45	0.44	0.39
This work	4.0	-	0.0175	★	★	0.39	0.32	0.32
0.083 m ID CFB	5.0	-	0.0175	★	★	0.38	0.30	0.23

★ Not applicable here, eq. (10) is used instead of eq. (4).

constant γ in equation (8) as calculated with equation (9) ($\gamma=0.259$) has been modified in a systematic way. For each gas distributor (thus each facility) one value of γ is estimated for all operating conditions (U_g and H_x) to meet the experimental c_{vx} and bubble dynamics. Thus the characteristics of each different gas distributor with respect to the initial bubble formation is accounted for.

Bubble dynamics are in agreement with the experiments only with low values of the visible bubble flow; $\varphi=0.0175$ gives good results. They can be expected at these high gas velocities although it should be pointed out that the low values of φ obtained should be seen as an indication of high through-flow of gas rather than exact values. Bubble diameters grow to the order of magnitude of the bottom bed height, which favours the temporally channelling or short cut of gas [6]. This is in agreement with visual observations of the bottom bed in the 1.2x0.8 m riser, which contains bubbles which have a considerable upward directed flow of solids.

For the 1.2x0.8 m riser it is found that $\gamma=0.325$ gives good results for all operating conditions (U_g and H_x). At the high gas velocities of this work it may be doubted that the assumption underlying equations (8) and (9) (i.e. each bubble cap acts as an individual bubble generator) is still valid. More than one bubble cap will contribute to an initial bubble formed. However, provided that γ is adjusted to meet the gas distributor characteristics, equation (8) may still be used to account for the dependency of the initial bubble size on the excess gas velocity U_g-U_{mf}.

A c_{vx} is calculated for the 12 MW$_{th}$ boiler with the same φ as used for the 1.2x0.8 m facility, see table 3.

Good model results for several superficial gas velocities are obtained when $\gamma=0.386$. The 12 MW$_{th}$ boiler has more bubble caps per unit bed surface (see table 1) and a lower gas distributor pressure drop and hence intial bubble formation is different. For the gas distributor of the 12 MW$_{th}$ boiler larger bubbles are calculated than for the 1.2x0.8 m riser at similar superficial gas velocities. This is in agreement with experimental observations at circulating conditions made in previous work [3].

Similarly, the bottom bed concentrations was modelled for the 0.083 m ID riser. Initial bubbles in this riser are of the same size as the riser diameter, so the slug rise velocity (equation (10)) has been used instead of the bubble rise velocity (equation (4)). Bottom bed concentrations are calculated which are considerably lower than those obtained in both large size facilities, which is in line with the trend in the experimental data, see table 3. Equation (10) is not capable to describe slug rise velocities at high superficial gas velocities and therefore we may expect the difference between the calculated and experimental bottom bed concentration at 5 m/s.

Figure 1 shows the calculated and the experimentally determined Kolmogorov entropy for the 1.2x0.8 m riser and 12 MW$_{th}$ boiler for runs at different superficial gas velocities. The proportionality constant ξ in equation (11b) was optimized for all operating conditions and facilities to be 8.9. Model results of Kolmogorov entropy are in agreement with experimental results for the different superficial gas velocities, different riser solids holdups (i.e. bottom bed heights H_x) and the two facilities. The N_{bubble} is in between the peak frequency in the spectrum and the cycle frequency.

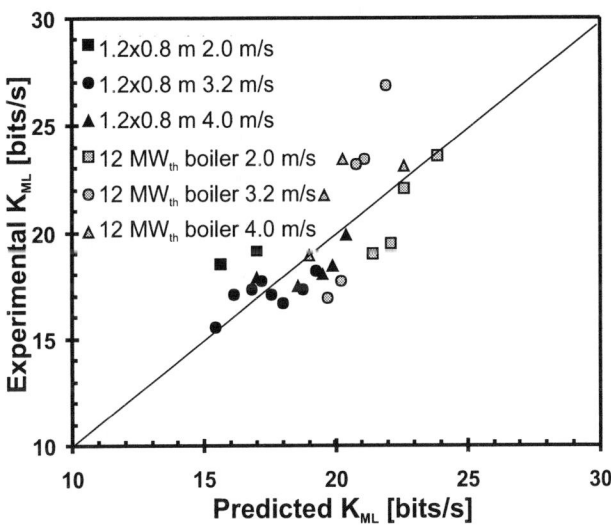

Figure 1 Experimental K_{ML} as a function of the predicted K_{ML}.

Model results of the dynamics of the bottom bed of the small size risers could not be made, because an adequate slug length relation at these high gas velocities is lacking.

Figure 2 shows the decay constant a obtained with equation (13) from experimental voidage profiles of the five CFB facilities in this work, see table 1. The decay constant a decreases when the gas velocity increases, in agreement with previous observations [4,5]. The decay constant a increases when the width or diameter of the riser increases, which is in contradiction with previous observations [4]. Furthermore, the decay constant a decreases significantly with the riser pressure drop (and thus with solids mass flux [10]) at constant U_g in the 0.083 and 0.12 m ID risers. For the 0.083 m ID riser a is calculated both with equation (13) and (15). The decay constant a calculated with equation (13) is higher than the one calculated with equation (15) for low riser pressure drop and the decrease of the decay constant a with an increase of riser pressure drop is found for both equation (13) and (15). For the 12 MW_{th} boiler and the 1.2x0.8 m riser equation (15) gives a poor fit of the axial solids profile above the bottom bed, since the transport zone is significantly present in these risers [1]. Therefore, it is preferred not to use equation (15) for these large risers.

CONCLUDING REMARKS

The axial solids distribution in large size risers and the dynamic characteristics of the bubbles present in the bottom bed can be described with simple models. The model simulations indicate that the major part (>95 %) of the gas passes the bed other than as visible bubble flow. The gas distributor plays an important role in initial bubble formation, which needs to be investigated in more detail. The decay constant a for the splash zone is significantly dependent on riser solids holdup in narrow risers and increases significantly with riser width.

ACKNOWLEDGEMENT

This work was supported by the Netherlands

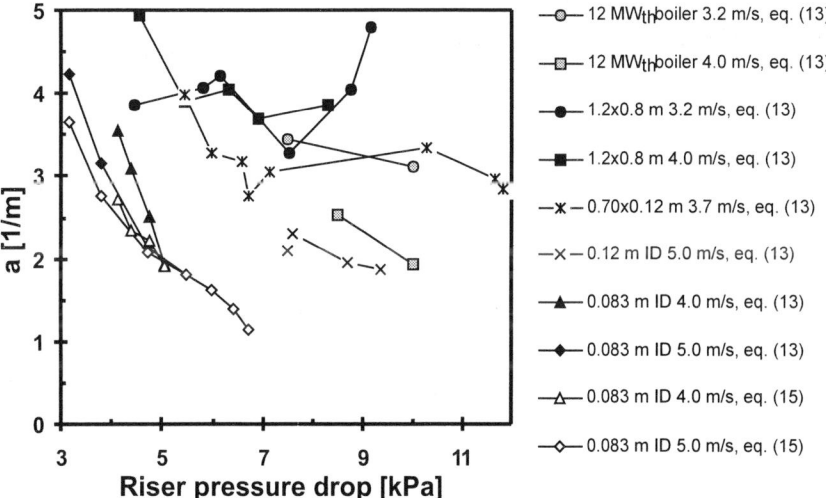

Figure 2 Decay constant a as a function of ΔP_{riser}.

Organisation for Scientific Research (NWO), which granted R.C. Zijerveld for his PhD study and by the Swedish National Board for Industrial Development (NUTEK). Their support is largely acknowledged.

NOTATION

a	decay constant, 1/m
A_{bed}	bed surface, m^2
c_v	solids volume concentration, -
c_{v2}	solids volume conc. from dispersed phase, -
c_{vd}	solids volume concentration dense phase, -
c_{vexit}	solids volume concentration at exit, -
c_{vmf}	solids volume concentration at U_{mf}, -
c_{vx}	solids vol. concentration bottom bed, -
c_v^*	solids volume concentration at saturated carrying capacity, -
d_v	bubble volume equivalent diameter, m
d_{v0}	initial d_v, m
d_p	solids mean diameter, m
d_t	equivalent bed diameter, m
g	gravity acceleration, m/s^2
G_s	solids mass flux, kg/(m^2 s)
H_{exit}	riser height, m
H_x	bottom bed height, m
K	decay constant transport zone, 1/m
K_{ML}	Kolmogorov entropy, bits/s
N_{Ar}	Archimedes number, -
N_{bubble}	number of bubble eruptions, 1/s
N_{nozzle}	# of distributor orifices, -
N_{Re}	Reynolds number $=(U_g-U_{mf})d_p\rho_g\mu_g$, -
U_b	bubble rise velocity, m/s
U_g	superficial gas velocity, m/s
U_{mf}	minimum fluidization velocity, m/s
U_{slug}	slug rise velocity, m/s
U_t	terminal velocity, m/s
Z	height above the dist. plate, m

Greek symbols

ΔP_{dist}	distributor pressure drop, kPa
ΔP_{riser}	riser pressure drop, kPa
ϵ_b	bubble volume fraction, -
γ	parameter defined in eq. 9, -
μ_g	gas viscosity, N*s/m^3
ξ	parameter defined in eq. 11, -
ϑ	parameter defined in eq. 6, -
ρ_g	gas density, kg/m^3
ρ_s	solids density, kg/m^3
φ	parameter defined in eq. 5, -

LITERATURE CITED

1. **Johnsson, F., A. Svensson and B. Leckner**, "Fluidization Regimes in Circulating Fluidized Bed Boilers," in *Fluidization VII*, O.P. Potter and D.J. Nicklin (eds.), Engineering Foundation, p. 471 (1992).

2. **Arena, U., A. Marzocchella, L. Massimilla and A. Malandrino**, "Hydrodynamics of Circulating Fluidized Beds with Risers of different Shape and Size," *Powder Technol.* **70** p. 237 (1992) and **71** p. 116 (1992).

3. **Zijerveld, R.C., F. Johnsson, A. Marzocchella, J.C. Schouten and C.M. van den Bleek**, "Fluidization Regimes and Transitions from Fixed Bed to Dilute Transport Flow", submitted to *Powder Techn.* (1996).

4. **Kunii, D and O. Levenspiel**, "Entrainment of Solids from Fluidized Beds I. Hold-up of Solids in the Freeboard II. Operation of Fast Fluidized Beds," *Powder Techn.* **61**, pp. 193 (1990).

5. **Johnsson, F. and B. Leckner**, "Vertical Distribution of Solids in a Circulating Fluidized Bed Furnace," in *Proc. of FBC 13, May 7-10, 1995, Orlando*, K.J. Heinschel (ed.), p. 671 (1995).

6. **Johnsson, F., S. Andersson and B. Leckner**, "Expansion of a Freely Bubbling Fluidized Bed," in *Powder Techn.* **68**, p. 117 (1991).

7. **Werther, J. and J. Wein**, "Expansion Behavior of Gas Fluidized Beds in the Turbulent Regime," *AIChE Symp. Ser.* **90**, 301, p. 31 (1994).

8. **Schouten, J.C., F. Takens and C.M. van den Bleek**, "Maximum-likelihood Estimation of the Entropy of an attractor," *Phys. Rev. E* **49**, p. 126 (1994).

9. **Schouten, J.C., M.L.M. vander Stappen and C.M. van den Bleek**, "Scale-up of Chaotic Fluidized Bed Hydrodynamics," *Chem. Eng. Sci.* **51**, p. 1991 (1996).

10. **Zijerveld, R.C., F. Johnsson, A. Marzocchella, J.C. Schouten and C.M. van den Bleek**, "Chaotic Hydrodynamics in the Riser Bottom Zone of Circulating Fluidized Beds of different Size," in *Preprints of 5th Int. Conf. on Circulating Fluidized Beds, Beijing (China), May 28- June 1*, DT2 (1996), accepted for the final proceedings (1997).

Micromechanics of Particle-Particle and Fluid-Particle Interactions in Flows of Particulate Systems

R. Ocone
Department of Chemical Engineering, University of Nottingham, Nottingham NG7 2RD, England

The behaviour of two-phase solid-fluid systems is determined essentially by the heterogeneous nature of the material considered. Treatment of the solid phase as a continuum relies greatly on statistical mechanics, which is not used simply to predict the value of the rheological parameters appearing in the constitutive equations, but to establish the very form of the constitutive equations themselves.

In this work attention is devoted to the micromechanics of fluid-particle systems. Attention is focused on the steps that, starting from the microstructural behaviour, lead to closure equations. The closure equations are shown to contain the local morphology, therefore, when performing actual calculations, it is not sufficient to keep track of the kinematics of motion and of the stress only. Some numerical results are presented to illustrate how they can be affected by the kind of particle-particle and fluid-particle interactions chosen.

Introduction

One of the main problems in modelling the behaviour of granular materials is to postulate the right constitutive equations. This is a difficult task: the description of some simple experiments, which can be performed with granular materials, shows that their mechanical behaviour is not of easy description. Statistical theories are often the tool to be used not only to predict the value of the rheological parameters appearing in the constitutive equations, but also to establish their form.

Statistical theories are based first of all on the identification of what are the microstructural elements considered in the theory. For granular materials the microstructural elements are the solid particles.

The logic of any statistical theory may be described as follows:

a) Identify the microstructural elements and their "local" morphology. By "local" one means in a neighbourhood of a point which is small enough to be regarded as a differential elements in the overall description, yet large enough to contain a statistically significant sample of the microstructural elements. By "morphology" we mean, in a somewhat loose sense, the average state of the microstructural elements within such a neighbourhood.

b) Find how the stress tensor can be expressed in terms of the local morphology and kinematics.

c) Write down a differential equation for the rate of evolution of the morphology under the action of the local kinematics of the main motion. In performing this step, one immediately realises why the mechanical response of granular material is sensitive to the flow type: the rate of evolution of the local morphology is such that the latter does depend on the flow type, and hence so does the stress tensor.

d) For any assigned kinematics of motion solve the obtained under c) above for the instantaneous morphology, and the equation under b) above for the instantaneous stress.

e) If possible (which is not often the case), eliminate the morphology between the equations so as to obtain directly a constitutive equation relating the instantaneous stress to the kinematics of motion.

When e) can indeed be performed, one in fact obtains a constitutive equation in the classical sense. When, as often is the case, step e) cannot be performed, d) above still yields the equivalent of a constitutive equation - with the additional complication that, in performing calculations, one needs to keep track not only of the kinematics of motion and of the stress, but also of the morphology.

Granular Materials
In the case of granular materials, the first part of the step a) is trivial: the microstructural elements are the grains themselves. Every particle may move with an instantaneous velocity u, and some sort of average velocity of the particulate phase, $v = <u>$, may be defined. The question of what one means by local average reserves subtle conceptual difficulties in the case of granular materials, since the ratio of the length scales of the overall description and of the microstructural elements is not large. Ideally one would want to define an average such that exact solution of the averaged equations of motion coincides with the averaged solution of the exact equations.

The average kinetic energy per unit mass of the particles, K, is $<u^2>/2$. Now, let c be the instantaneous oscillation velocity of the particles, $u = v + c$. Since $<c> = 0$, $K = <v^2>/2 + <c^2>/2$, and hence, in addition to the kinetic energy of the main motion, $<v^2>/2$, one has a "fluctuating" energy $<c^2>/2$. This has been called the "pseudo-temperature", T, by analogy with the classical kinetic theory of gases (with grains playing the role that molecules play in the kinetic theory). Step a) is now completed: **the "local morphology" is simply T.**

Step b) is accomplished by developing a statistical theory which bears close analogy with the kinetic theory of gases, with the individual grains playing the role that molecules paly in the classical statistical theory. The result is a constitutive equation of a compressible, Newtonian fluid, with, however, the parameters depending on both T and the local solid volume fraction, v. This shows how crucial it is to perform step c), since the equations of the motion are not closed unless one can calculate the spatial and temporal distribution of T.

Step c) implies writing down a balance equation for the fluctuating energy. When the latter is regarded as T itself, the conceptual content of such a balance equation is not easy to grasp. If one regards the fluctuating energy as a unique function of T the balance of fluctuating energy reveals its strong analogy with the first law of thermodynamics (Astarita and Ocone, 1994). However, inelasticity of particle-particle collisions introduces a substantially new element: a granular material cannot be sustained in a "thermalised" state (a non-zero T), unless mechanical energy is continuously supplied to it so as to balance the fluctuating energy loss due to inelastic collisions.

Step d) turns to be rather difficult, except for the simplest flow geometries, particularly because there are problems involved with writing down the appropriate boundary conditions. Furthermore, the two-phase nature of the system considered plays a crucial role, and a momentum balance for the interstitial fluid phase needs to be written down as well; this is coupled with the momentum balance for the particulate phase by way of an interaction.

Step e) cannot be performed since the variable T can never be eliminated from the governing equations.

Some examples (how the morphology affects the constitutive equations and results)

Let's start considering a very simple illuminating example of how the morphology cannot be eliminated from the governing equations: steady shear flow between parallel plates at distance H, with the upper plate in motion with velocity V. From statistical theory step b) above furnishes a constitutive equation of a compressible Newtonian fluid, therefore, the tangential force per unit area in the direction of motion, τ, is:

$$\tau = \frac{\mu V}{H} \quad (1)$$

But the viscosity, μ, depends now on the morphology, ie. on T. T itself is established by the following requirement: the rate of energy input per unit volume due to the macroscopic shear, $\mu(V/H)^2$, balances the rate of dissipation due to inelasticity of collisions, I. A simple dimensional argument shows that I is proportional to $T^{3/2}$. It follows that T scales with V^2, and hence μ with V. This explains why τ turns to be proportional to the square of V.

If one runs experiments at different values of H, there seems to be two possible behaviours, depending on the physical nature of the surfaces of the two plates. The first kind of behaviour is again that τ is a unique function of V/H, say:

$$\tau = K \left(\frac{V}{H}\right)^2 \quad (2)$$

The second kind of behaviour is more complex. τ turns to be proportional to the square of V at sufficiently high values of V, no matter what H may be. However, at low values of H, the previous results start holding at comparatively low values of V; conversely, at high values of H, Eq. (1) holds up to very large values of V. In other words, if particle-wall collisions are elastic τ is always proportional to the square of the ratio V/H. On the other hand, if particle-wall collisions are inelastic τ is proportional to the square of v: morphology and micro mechanics can affect significantly the experimental results one can obtain in a shear flow experiment (Ocone and Astarita, 1994).

The previous result is obtained for dry granular materials; when the interstitial fluid is taken into account the situation may become even more complex.

A very immediate example is the effect of the air drag on the micro mechanics of granular materials. Depending on the concentration of solid and on the kind of flow (laminar or turbulent), not only the constitutive equations depend on the morphology, but even the balance equations, as for instance the "pseudo-energy" balance. In fact in some circumstances it is plausible that dissipation of pseudo-energy is not due only to inelastic collisions but also to the gas drag against the fluctuating velocity of particles. In Figure 1 calculations are reported which include in the balance of pseudo-energy a term which takes into account fluctuation related dissipation. If such a term is not incorporated into the pseudo-thermal energy balance, the gas and solid velocity profiles obtained are quite parabolic (see Figure 2); on the other hand, if the fluctuation related dissipation is considered the profiles are quite flat (see Figure 1).

Gidaspow, 1990). In particular, the dependence of the drag force on the relative velocity has been considered as linear and square. Letizia et al. (1997) have presented calculations which show how the choice of

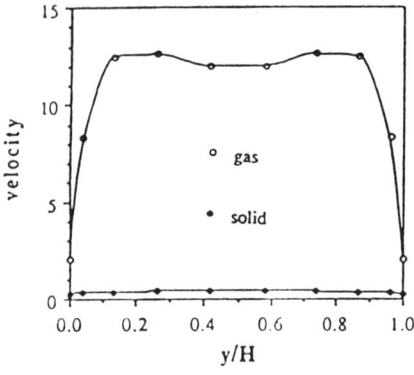

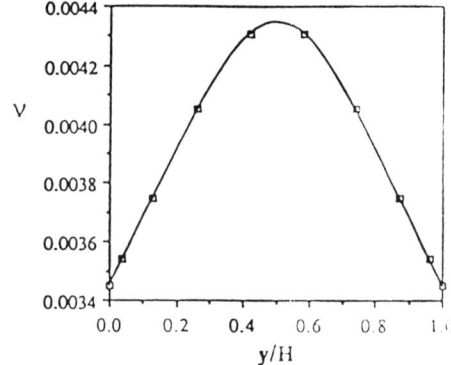

Figure 1. Velocities and solid volume fraction distributions including the damping effect

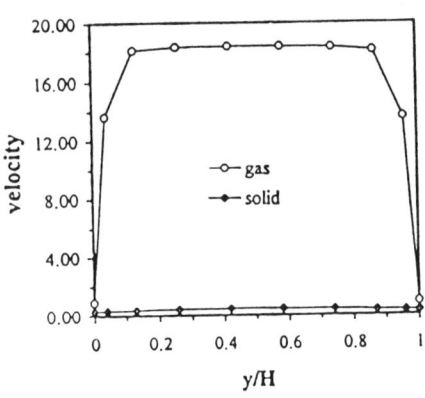

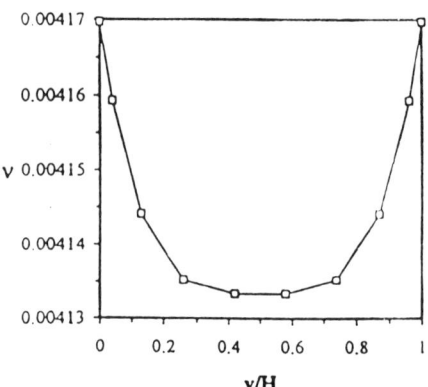

Figure 2. Velocities and solid volume fraction distributions without the damping effect

The drag force that the interstitial fluid exerts on the particle phase is expressed in different forms by different authors (e.g. Sinclair and Jackson, 1993; Ding and

the different dependence can affect the velocity and solid volume fraction profiles. Many other effects can be considered which show how the micro mechanics of granular materials can play an important role in determining the flow behaviour. It has been recently shown (Letizia et al., 1997) that particle-particle interactions and gas-particle interactions can have a very strong effect on the behaviour of granular materials in flow. In particular, some observed flow recirculations and related changes in flow regimes in circulating fluidised beds (Bodelin et al., 1993) have been shown to be related to the micro mechanics of particle-wall interactions (Letizia et al., 1997). Results are shown in Figure 3 for different particle-wall kind of interaction.

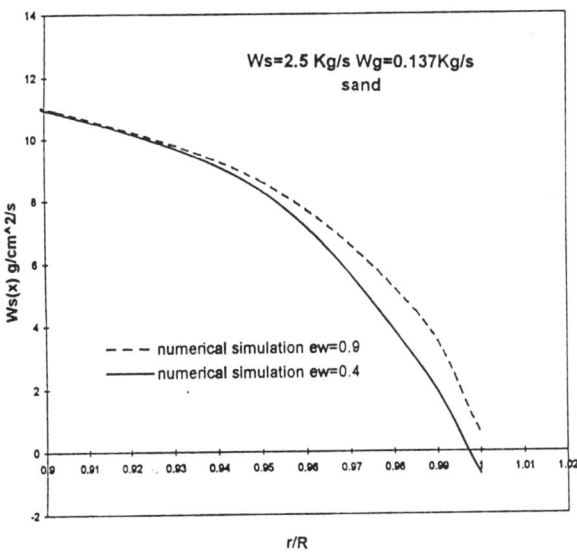

Figure 3. Local flux profiles near the wall at different values of particle-wall coefficient of restitution, ew.

There are in the literature many other examples which confirm how the morphology cannot be taken out from the constitutive equations. Cody et al. (1996), using a new non-intrusive probe make measurements of the granular temperature, T, in fluidised beds. They observe a quite dramatic change of the dependence of T on the superficial gas velocity when the Geldart A/B boundary is crossed; this validates the idea that the different fluidization behaviour of A and B particles lies in a difference in gas particle dynamics.

The interstitial fluid nature is another example of the complexity and morphology depending behaviour of granular materials in flow. In all the cases reported above, the interstitial fluid is considered "almost" inviscid (inertial limit) influences the particle phase stress tensor (the only action of the fluid on the solid phase is through the drag force). However, there are analysis (Jenkins and McTigue, 1990; Ocone and Astarita, 1995) considering the interstitial fluid having a very high viscosity (viscous limit). In this limit the stress tensor is dominated by lubrication theory type of stresses arising from the squeeze flow of the fluid between two close particles in relative motions toward each other. Again, if one develops constitutive equations for the inertial and viscous limits, one ends up with two different constitutive equations with a different dependence on T, then on the morphology.

Conclusions

Granular materials are very complex "fluids" and it is practically impossible to develop constitutive equations which describe the rheology of such materials in all possible flow fields. Statistical theories try to overcome this problem, but unfortunately the morphology cannot be easily and always eliminated from the constitutive equations. The consequence is that, even if one relies on the validity of the constitutive equations obtained, one cannot use them in all flow fields. Macroscopic experimental results can be different for different morphologies, and the only way to have a clear understanding of the flow behaviour of granular materials is to perform more experiments aimed to gain information on their microstructure and micromechanics.

References

Astarita, G. and R. Ocone, "Large Scale Statistical Thermodynamics and Wave Propagation in Granular Flow", I&EC Research, **33**, 2280, 1994

Bodelin, P., Y. Molodtsof, A. Delebarre, "Flow Structure Investigations in a CFB", in Proc. CFB-4, ed A. Avidan, New York 1993

Cody, G.D., D.J. Goldfarb, G.V. Storch, A.N. Norris, "Particle Granular Temperature in Gas Fluidized Beds", Powd. Tech., **87**, 211, 1996

Ding, J. and D. Gidaspow, "A Bubbling Fluidization Model using Kinetic Theory of Granular Flow", AIChE J., 36, 523, 1990

Jenkins, J. T., and D. F. McTigue, "Transport Processes in Concentrated Suspensions: the Role of Particle Fluctuations", pp 70-79, in D. D. Joseph and D. J. Schaeffer (eds), "Two Phase Flow and Waves", Springer Verlag, New York 1990

Letizia, L., A. Delebarre, R. Ocone, "Upflow of Solid/Gas Suspensions; the Role of Particle-Particle and Particle-Gas Interactions", AIChemE Research Event, Nottingham 1997

Ocone, R. and G. Astarita, "Shear Flow of Dry Granular Materials", Indian Chem. Engr., Section A, **36**, 151, 1994

Ocone, R. and G. Astarita, "Grain Inertia and Fluid Viscosity dominated Granular Flow Rheology: a Thermodynamic Analysis", Rheol. Acta, **34**, 323, 1995

Sinclair, J.L., and R. Jackson, "Gas-Particle Flow in a Vertical Pipe with Particle-Particle Intercations", AIChE J., **35**, 1473, 1989

Magnetic Resonance Imaging of Granular Convection

James B. Knight, Heinrich M. Jaeger, and Sidney R. Nagel
The James Franck Institute and Department of Physics, The University of Chicago, Chicago, IL

We describe experiments that probe the detailed nature of convection patterns occuring in dry, cohensionless granular material subjected to vibrations. For cylindrical containers filled with poppy seeds magnetic resonance imaging (MRI) is used to map the local convection velocity throughout the interior of the system in a noninvasive manner and with high resolution. Using MRI, we also show how granular convection is able to induce size separation in vibrated mixtures of two particle sizes.

Dry, cohesionless granular material convects in response to applied vertical oscillations. For peak vibratory accelerations less than that of gravity, no relative particle motion can be sustained; above this threshold, particles circulate in a pattern dependent upon container geometry. In a container with rough, vertical walls, flow is upward in the center and downward in a thin layer along the walls.

Modern scientific interest in such granular convection, which was originally reported by Faraday in 1831 [1], stems from its utility in the study of granular flow and from its implication in the industrially relavent problem of size separation [2]. Unlike many examples of granular motion, granular convection is stable over long periods of time and reproducible. The flow is circulatory, and the particles are confined to a closed volume, eliminating the need to continually add material for experiments of long duration. By varying the peak acceleration or frequency of the driving vibration, the convection velocity can be changed by orders of magnitude, permitting study of a wide range of granular flow speeds in a single system. These qualities, coupled with the oppportunity for comparison with experiment, have also prompted extensive theoretical and simulational work [3].

Large particles rise to the top of a vertically shaken granular mixture. Commonly referred to as the 'Brazil Nut' problem, size separation can be devastating to industrial processes. Convection in containers with rough, vertical walls is an important mechanism for this phenomenon: large particles rise upwards in the broad central flow, but remain at the surface if they are too big to enter the thin downward flow [4]. By controlling convection, this process can be halted and even reversed [4,5].

Experimental study of granular convection, and of granular flow in general, is complicated by the opacity of granular material. In the past, this has limited experimental observation to external features of three dimensional flow or to two dimensional systems. Tracer particles and low-resolution imaging techniques have been used with three dimensional flow, but without the precision necessary to establish experimental 'benchmarks' against which theory and simulations can be tested [6]. Recently, however, magnetic resonance imaging (MRI) has been applied as a high-resolution, noninvasive probe of granular flow [7,8,9]. In this paper, we briefly review the technique and describe some of the insights it has provided into granular convection. A detailed treatment can be found in References [8] and [9], on which this paper is based.

Magnetic resonance imaging is ideally suited to liquids, but Nakagawa *et al.* [7] have shown that oil bearing seeds can contain enough free protons in the liquid state to produce a detectable magnetic resonance signal. We use white poppy seeds for their high oil content, dry exterior, and small size (1 mm)[10]. In a typical experiment, a cylinder is filled with white poppy seeds and placed on a nonmetallic vibrating platform within the bore of a GE/Bruker 4.7T MRI magnet. A layer of seeds is epoxied to the walls of the cylinder to control friction. All of the components within the magnet bore must be nonmetallic, and the sample platform is

---

The University of Chicago, Chicago, Illinois. J.B. Knight is presently at Princeton University, Princeton, New Jersey.

coupled through a long, rigid rod to an electromagnetic vibration exciter placed 3m from the magnet. The mechanical limitations of this shaking apparatus restrict shaking to discrete 'taps' - individual sinusoidal oscillations separated by a waiting period sufficient for particle movement to cease and to acquire the necessary magnetic resonance data [11]. We parameterize the strength of the applied acceleration by Γ, the dimensionless ratio of the applied peak acceleration to that of gravity.

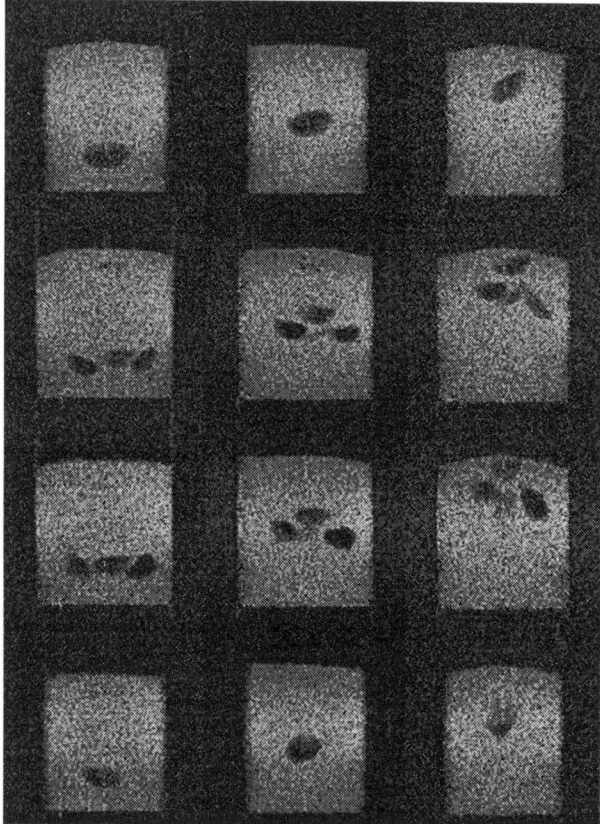

Figure 1. Magnetic resonance images showing the rise of five coffee beans in a bed of poppy seeds. The four rows correspond to vertical slices through the cylinder from back to front; the columns show the time progresion, with time increasing from left to right.

Figure 1 illustrates convective size separation with magnetic resonance images of a cylindrical container filled with poppy seeds. The bright areas correspond to areas of high oil concentration, and thus high signal, while the large dark areas within the bulk are coffee beans. The beans act as tracer particles and are transported upwards by the convective flow. Time in Figure1 runs from left to right; the initial placement of the beans is visible in the left column of images, and the next columns show images at later times. The four rows show four vertical cuts through the cylindrical container: one close to the front, two near the center (which are slightly wider), and one closer to the back. The container was tapped and then held stationary as each image was collected. The beans clearly rise as the container is shaken. Once the beans reach the top of the pile, they remain on top, as their diameter exceeds the width of the downward flowing region.

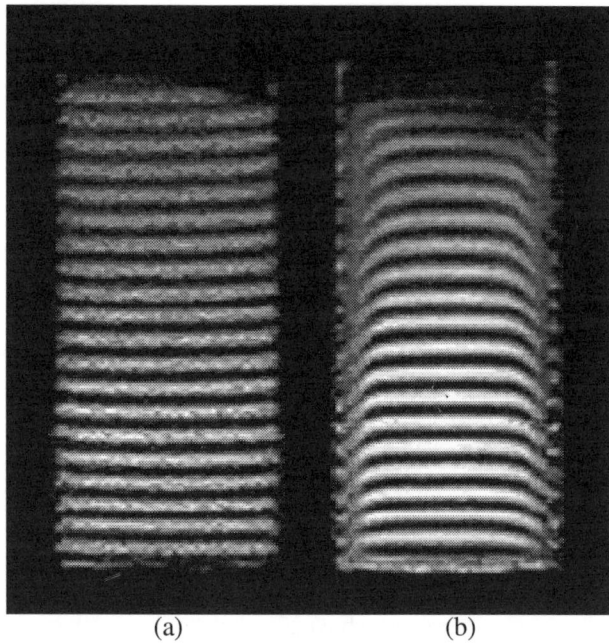

Figure 2. (a) Magnetic resonance image of a 3mm slice through the center of an unshaken acrylic cylinder 25 mm in diameter. (b) Magnetic resonance image showing the deformation of the spin-tagging pattern after one tap with $\Gamma = 6$.

Large tracer particles are a relatively low resolution, inefficient means of measuring flow. Spin-tagging offers a more direct approach [8,9,12,13]. Layers of seeds can be magnetically 'tagged' by modulating the longitudinal spin-polarization in the vertical direction. Figure 2a shows a control image of a slice through the center of a cylinder filled with poppy seeds taken after spin-tagging but without shaking. The bright areas in the image correspond to the maxima of the spin modulation. Flow translates the tagged particles, distorting the initially horizontal stripes. Figure 2b is an image of the deformation after a single shake with $\Gamma = 6$. The stripes have bent in a manner consistent with particle flow upward in the center of the container and downward along the sides. The layer of seeds glued to the cylinder wall has not moved, and serves as a reference point for the initial position of the tagged layers. Because of the rapid decay of the spin-tagging pattern due to thermal randomization of the spins [14], 256 identical taps are necessary to obtain a single image of the resolution shown in Figure 2. Furthermore, Figure 2b is an average of 8 such images, taken consecutively. The clarity of the image illustrates the stability and reproducibility of convective flow.

Figures 3a-d are magnetic resonance images showing the original spin tagging pattern and the deformation after a single tap for several values of the peak vibration acceleration. Flow speeds are highest near the top surface and decrease rapidly with depth; it has been shown using tracer particles and MRI data that this decay is exponential [8,9]. Near the edges of the flow, the data is noisy, particularly at the higher accelerations. This is due to diffusive spreading of the tagged particles, and provides a measure of particle diffusion [13].

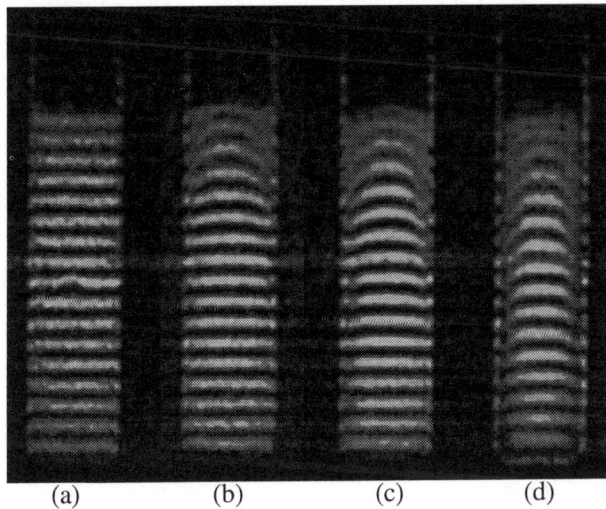

(a) (b) (c) (d)

Figure 3. Magnetic resonance images of a 3mm slice through the center of an acrylic cylinder (12.7mm in diameter) showing the spin-tagging pattern (a) before tapping and after one tap of (b) $\Gamma=4$, (c) $\Gamma=6$, and (d) $\Gamma=8$.

Images such as Figures 2 and 3 have been analyzed digitally to precisely measure the convective flow velocity as a function of both the depth and the radial coordinate. The original tagging pattern is sinusoidal, and the peak positions before and after shaking can be determined to within fractions of a particle diameter. Figure 4 compares the velocity at the same depth and acceleration in three different cylindrical containers. The curves are offset for clarity, and a dotted line is included with each that identifies zero velocity and spans the container width. Flow above the dotted line is upward; flow below is downward. The individual velocity curves are relatively flat in the center of the container and decrease rapidly near the walls. This radial dependence is captured equally well by fits to a cosh and a modified Bessel function of order zero, with the key feature being an exponential change in the velocity away from the walls [9]. Fits to a parabolic curvature, which arises in the simple laminar flow of a liquid through a pipe, fail to capture the flat central flow.

The width of the downward flowing region, w_d, is an important parameter in convective size separation. In Figure 4, it is the distance from the wall of the container to the point at which the flow changes direction. Within the range of data studied, this width is independent of both acceleration and depth [9]. As is

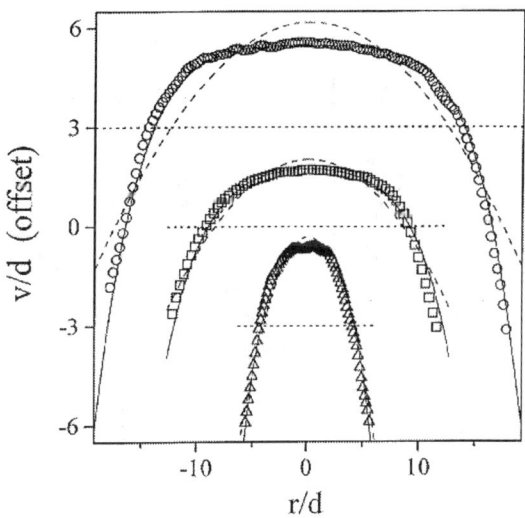

Figure 4. Three representative velocity profiles obtained from MRI pictures such as the one in the previous figures. Shown are the radial velocities $v(r)$ for three container diameters at fixed depth into the container and applied acceleration $\Gamma=6$; the curves are offset for clarity. The dotted lines indicate the $v=0$ level and their width corresponds to the diameter of the container. The solid lines are fits to an exponential dependence, $v(r) \propto I_0(r/r_0)$, where I_0 is a modified Bessel function and r_0 a characteristic length that depends mainly on the container diameter and only very weakly, if at all, on the applied acceleration. The dashed lines show a parabolic profile for comparison. All values are normalized by the bead diameter, d. For details see Reference [9].

apparent from Figure 4, however, w_d increases with container diameter; detailed measurements show that the width of the downward flow is proportional to the container size. Container size is therefore a parameter which can be varied to control convective size separation.

Magnetic resonance provides a high-resolution, noninvasive technique for imaging three dimensional granular flow. Spin tagging further enables precision measurements of velocities within the flow. In this paper, we have briefly surveyed some of the results obtained with MRI for granular convection. The flow velocity decays exponentially with depth from the top surface. The radial dependence is captured equally well by fits to a hyperbolic cosine or a modified Bessel function of order zero, with both functional forms sharing the key feature of varying exponentially near the container walls. The width of the downward flow is proportional to container size, suggesting a practical control parameter for size separation. Simple, heuristic models in agreement with best fits to the data are presented elsewhere [8,9].

ACKNOWLEDGEMENTS

This work was supported by the NSF through MRSEC Grant DMR-9400379. We acknowledge additional support through DOE Grant DE-FG02-92ER25119 and from the David and Lucile Packard Foundation.

NOTATION

d   grain diameter
Γ   dimensionless acceleration parameter (peak acceleration applied during a vibration cycle normalized by the earth's gravitational acceleration, 9.81m/s^2)
r   radial coordinate, measured with respect to central container axis
v(r)   convection velocity, i.e., net grain displacement per vibration cycle, as a function of radial coordinate
w_d width of the downward flowing region near the container walls

LITERATURE CITED

1. Faraday, M., "On a Peculiar Class of Acoustical Figures; And on Certain Forms Assumed by Groups of Particles Upon Vibrating Elastic Surfaces," *Phil. Trans. R. Soc. London*, **52**, pp. 299 (1831).

2. For reviews of granular convection and the physics of granular materials in general see, e.g., Jaeger, H.M., and S.R. Nagel, "Physics of the Granular State," *Science*, **255**, pp. 1523 (1992); Behringer, R.P., "The Dynamics of Flowing Sand," *Nonlinear Science Today*, **3**, pp. 1-15 (1993); Hayakawa,H., H. Nishimori, S. Sasa, and Y.-H. Taguchi, "Dynamics of Granular Matter," *Japn. J. Appl. Phys.*, **34**, pp. 397 (1995); Jaeger, H.M., S.R. Nagel, and R.P. Behringer, "Granular Solids, Liquids, and Gases," *Rev. Mod. Phys.*, **68**, pp. 1259-1273 (Oct. 1996).

3. For a comprehensive list of work in the field, see references cited in [2].

4. Knight, J.B., H.M. Jaeger, and S.R. Nagel, "Vibration-Induced Size Separation in Granular Media: The Convection Connection,"*Phys. Rev. Lett.*, **70**, pp. 3728 (1993).

5. Knight, J.B., "External boundaries and internal shear bands in granular convection," *Phys. Rev. E*, in press (1997).

6. Ratkai, G. "Particle Flow and Mixing in Vertically Vibrated Beds," *Powder Technol.*, **15**, pp. 187 (1976); Harwood, C.F., "Powder Segregation Due to Vibration,"*ibid.* **16**, pp. 51 (1977); Baxter, G.W., R.P. Behringer, T. Fagert, and G.A. Johnson, "Pattern Formation in Flowing Sand," *Phys. Rev. Lett.*, **62**, pp. 2825 (1989).

7. Nakagawa, M., S.A. Altobelli, A. Caprihan, E. Fukushima, and E.-K. Jeong, "Non-invasive measurements of granular flows by magnetic resonance imaging," *Exp. Fluids*, **54**, pp. 16 (1993).

8. Ehrichs, E.E., H.M. Jaeger, G.S. Karczmar, J.B. Knight, V.Yu. Kuperman, and S.R. Nagel, "Granular Convection Observed by Magnetic Resonance Imaging,"*Science*, **267**, pp. 1632 (1995).

9. Knight, J.B., E.E. Ehrichs, V.Yu. Kuperman, J.K. Flint, H.M. Jaeger, and S.R. Nagel, "Experimental Study of Granular Convection," *Phys. Rev. E*, **54**, pp. 5726 (1996).

10. The poppy seeds are elliptical in shape, with a major axis of about 1mm and a minor axis of 0.7 to 0.8 mm.

11. Comparison with tracer particle data has confirmed that continuous shaking and tapping produce qualitatively similar convection (see Reference [9]).

12. Axel, L., and L. Dougherty, *Radiology*, **171**, pp. 841 (1989).

13. Kuperman, V. Yu., E.E. Ehrichs, H.M. Jaeger, and G.S. Karczmar, "A New Technique for Differentiating Between Diffusion and Flow in Granular Media using Magnetic Resonance Imaging," *Rev. Sci. Instr.*, **66**, pp. 4350 (1995).

14. The spin-lattice relaxation time, T_1, is about 200ms for poppy seeds in a 2T field.

Modelling of Particle Interaction Laws in Slow Shearing Granular Flows

D.M. Heyes
Department of Chemistry, University of Surrey, Guildford, GU2 5XH, England

U. Tüzün and J. Baxter
Department of Chemical & Process Engineering, University of Surrey, Guildford, GU 2 5XH, England

We show that in order to obtain realistic behaviour in Discrete Element modelling of slow shearing flows of granular materials it is important to adopt a sufficiently long-ranged interaction force law that allows a highly extended degree of rotational coupling between adjacent granules. This enables the granular assembly to consolidate and evolve microstructurally without requiring significant center of mass translational motion of the particles. Illustrative examples are provided from recent simulations carried out on model hoppers and in the formation of free-standing heaps by vertical feed.

Granular dynamics is a discrete element, DE, simulation technique that can produce a representation of a granular system as it evolves in time and space. The technique has been applied to most situations in which granular materials are handled (*e.g.*, hoppers [1][2], and chutes [3]). The most important component of any DE application is the choice of interaction law between the elements. The ability of the model to reproduce the behavior of the real material is a direct consequence of the ability of the interaction law to account for the physical processes operating on the most appropriate length and time scale for the assembly mechanics. The normal interaction laws used in simulations of granular materials have traditionally been based on single particle contact laws (e.g. Hertz [4]) or even simpler analytic forms ('springs and dashpots'). For example, consider two elastic spheres of undeformed diameter σ and mass m then the Hertz force law is,

$$F_{Hertz} = \begin{cases} (12E'/3\pi)\rho^{2/3}m^{2/3}g(r/\sigma), & r/\sigma \leq 1 \\ 0, & r/\sigma > 1 \end{cases} \quad (1)$$

*Department of Chemistry, University of Surrey, Guildford, GU2 5XH, UK.
†Department of Chemical & Process Engineering, University of Surrey, Guildford, GU2 5XH, UK.

where E' is an elastic modulus, ρ and m are the density and mass of the sphere and $g(x) = (1-x)^{2/3}$. For a Hooke's law interaction with force constant k we have,

$$F_{Hooke} = \begin{cases} (6k/\pi)\rho^{1/3}m^{1/3}h(r/\sigma), & r/\sigma \leq 1 \\ 0, & r/\sigma > 1 \end{cases} \quad (2)$$

where $h(x) = (1-x)$. While adequate for elastic spheres, these interactions have serious limitations when it comes to representing typical granular materials. The interaction force is set to zero at $r = \sigma$ and for greater separations which therefore does not allow for assembly connectivity that may arise from shape variation, surface roughness and polydispersity for example. Also in these interactions the force between the particles scales with arbitrary powers of the mass. For a uniform affine dilation or compression of the particles (*i.e.*, constant fractional change in the value of r/σ irrespective of the value of σ) the forces between the particles will scale with the mass to some power. Therefore larger particles with the same fractional contraction will exert a larger force on the surrounding assembly. While this might be satisfactory for highly compacted powders, in many processes involving powders (such as conveying, silo discharge and heap formation) different physical processes are at work where the normal loads are typically rather small but the plastic deformation of the material on the particle scale and above is rather large. The above interaction laws which ensure miniscule particle and assembly deformation for realistic material elastic constants will give rise to a misleading picture in these situations. We suggest that this issue has to be looked at again as it is the deformation of the bulk assembly that governs the mechanics of slow flowing granular systems, and that the appropriate interaction laws should scale with the assembly microstructure rather than with the contact (*i.e.*, asperity contact) dimensions and time scale (*i.e.*, microslip). The frequent observation that assembly macromechanics is insensitive to the material constants of the constituent granules already suggests as much. DE simulations we have carried out for hoppers and stockpile formation over recent years support this proposition. To summarise, interaction laws that act on the scale of the particle and the physical processes that occur on this scale rather than on the contact are required.

In any real granular material, bulk interactions between particles are rather poorly characterised, owing to uncertainties, associated with, for example, particle shape, surface roughness and polydispersity. The consequence of this is that if we wished to adopted a *first principles* model each granule in the assembly would have to be treated with a separate rather detailed interaction law, which even for a single pair would be difficult to evaluate. This is clearly unrealistic in the near future for computational reasons. Only relatively simple interaction laws can be used at present to permit real-time events to be modeled on a routine basis. The challenge has been to employ interaction laws that are economical to implement but nevertheless introduce the appropriate physics on the assembly lengthscale. Our solution to this problem has been to develop a family of mean-field interaction laws that are spherically symmetric but can generate highly cooperative assembly dynamics, necessary to produce slow shear flows through dilation. (Unlike a liquid, a granular material can **only** flow by first dilating, which is a highly cooperative process.)

By a process of continual refinement in recent years we have been developing interaction laws that generate increasingly realistic assembly dynamics. These interaction laws have two distinct regions. At short range they become highly repulsive to prevent significant particle overlap, and thereby ensuring high incompressibility of the assembly. However, at long-range (greater than the nominal particle diameter) they have a slowly decaying 'tail' that although comparitively weak compared with gravitational forces is essential because it can induce highly coupled particle motion in the assembly, especially through particle rotation. One analytic form for the normal interaction law, which we have called the Continuous Interaction, CI, force law is the soft-sphere interaction,

$$F_{CI} = \begin{cases} mgf(r/\sigma), & r \leq r_o \\ 0, & r \geq r_o \end{cases} \quad (3)$$

where

$$f(x) = (\sigma/r)^{n+1} = 1 + (n+1)(1-\frac{r}{\sigma})$$
$$+ \frac{(n+1)(n+2)}{2!}(1-\frac{r}{\sigma})^2 + \cdots \quad (4)$$

r_o is a truncation separation which is typically 10 to 20 % greater than the nominal particle diameter. We have maintained a mass dependent coefficient (*i.e.*, mg) in order to set the scale for contact deformation under gravity. This ensures that at rest under gravity the separation between two model granules positioned one vertically above the other is σ. The significance of the interaction in Equation (3) for separations greater than the nominal particle diameter is that it aims to represent some aspects of the deformation of the assembly with an extended region of low load engagement.

We argue that contact interaction laws in such models should be established and verified on the basis of the assembly mechanics (*i.e.*, behavior on the scale above that of the individual granule) rather than from the microcontact mechanics of individual contacts. In slow shearing granular flows, in particular, the interaction laws are dominated by geometrical issues associated with prolonged particle engagement within the assembly. It therefore would be unlikely that adequate interaction laws of use in DE simulation of dense phase granular materials could be derived solely by considering two granules alone, without considering the fact that they are embedded in a matrix of other granules. The effect of this is that the interactions between the particles that should be used are rather soft (to allow for a granule to interact with many of its neighbors at the same time) and do not necessarily correlate well with the material constants of the particles themselves. This is borne out by experiment by the relative insensitivity of wall stresses and discharge rates in hoppers on the material constants (e.g., elastic moduli) describing the individual particles. The fact that previous DE granular dynamics simulations have used unrealistically low value for the elastic constants in Equations (1) and (2) is essentially an admission of this fact. We argue that assembly deformation is what governs the mechanics of slow flowing granular systems, and that the appropriate interaction laws should scale with the assembly microstructure rather than the particle contacts which has been widely assumed to date.

There are other important forces that must be considered in the DE model and these will be discussed below for two dimensional systems, although extensions to 3D have been made [2].

NORMAL INTERACTIONS

An important function of the normal force between granules i and j, say, is to keep the particles from overlapping unrealistically and also provide more long range weaker interaction which strongly influences the ability of the particle to undergo relative displacements in the perpendicular or 'tangential' direction. The total force, $\underline{F}_{ij}$, is the sum of these normal, $\underline{F}_{ij}^N$ and tangential $\underline{F}_{ij}^T$ components.

$$\underline{F}_{ij} = \underline{F}_{ij}^N + \underline{F}_{ij}^T \qquad (5)$$

Therefore the total force on granule i is

$$\underline{F}_i = \sum_{k=1}^{n_c} \underline{F}_{ik} + \underline{F}_i^g \qquad (6)$$

where the sum is over all the contacts, n_c and $\underline{F}_i^g$ is the vertical (y−direction) gravitational force acting on granule i.

Both the normal and tangential force consists of a term that depends on position or accumulated displacement, and a velocity dependent force representing contact damping. For the normal direction we have,

$$\underline{F}_{ij}^N(r_{ij}, \dot{\underline{r}}_{ij}) = \left(F_e^N(r_{ij}) - C_N(r_{ij})\dot{\underline{r}}_{ij}.\hat{\underline{n}} \right) \hat{\underline{n}} \qquad (7)$$

where in $3D$, $\hat{\underline{n}} = (r_{ij}.\underline{x}_{ij}, r_{ij}.\underline{y}_{ij}, r_{ij}.\underline{z}_{ij})/r_{ij}$ is the normal unit vector for the contact. Kinetic energy is lost by the particles through relative motion of the contact surfaces. This is introduced by including a velocity-dependent dissipative force which brings into the model the inelastic nature of real contact interactions. The damping law is based on the solution of a classical damped harmonic oscillator, with force constant k. For the normal interaction, the damping force is the product of a damping coefficient, $C_N(r)$, times the instantaneous relative velocity component in the normal direction. $C_N(r)$ was chosen to be an arbitrary fraction (50 %) of the critical damping coefficient, $C_c = 2(mk(r))^{1/2}$, where $k(r) = -dF_e^N(r)/dr$.

SLIDING INTERACTIONS

Similarly in the sliding or 'tangential' direction the force has displacement and relative velocity components,

$$\underline{F}_{ij}^T(\delta_F, \dot{\underline{r}}_{ij}) = \underline{F}_e^T(\delta_F) - C_T(\delta_F)(\dot{\underline{r}}_{ij}.\hat{\underline{t}})\hat{\underline{t}} \qquad (8)$$

where $\hat{\underline{t}} = (r_{ij}.\underline{y}_{ij}, -r_{ij}.\underline{x}_{ij})/r_{ij}^2$ is the tangential unit vector. The first term on the righthandside of Equation (8) depends on an accumulated contact sliding deformation, δ_F, which we have chosen the form,

$$\underline{F}_e^T(\delta_F) = \mu F_e^N \left(1 - (1 - \frac{|\delta_F|}{\delta_{max}})^{3/2} \right) \hat{\underline{t}} \qquad (9)$$

where μ is an 'effective' friction coefficient (we have chosen typically $\mu \sim 0.4 - 0.6$). The tangential force has an initial elastic response for small values of δ_F merging into a plastic contact deformation force law for larger δ_F. As $\delta_F \rightarrow \delta_{max}$ the compliance increases and the frictional force tends to Amonton's Law i.e., $F_e^T = \mu F_N$. The contact displacement, δ_F, is a cumulative scalar quantity representing the distortion of the local microstructure arising from the relative movement of two contacting particles caused by a combination of rolling and sliding, with the form

$$\delta_F = \int_0^t [\dot{\underline{r}}_{ij}(s).\hat{\underline{t}}(s) - (\dot{\theta}_i(s) + \dot{\theta}_j(s))r_{ij}(s)]ds \qquad (10)$$

where the time variable $s = 0$ is when the two particles first come into contact or 'engage' at a separation $r_{ij} = r_o$. The normal deformation is $\delta_N = r_o - r > 0$. In many of our simulations we have chosen the values $r_o = 1.2\sigma$, which corresponds to a cut-off force of 10^{-3} mg for the $n = 36$ force law.

Although a Hertzian analysis for normal interactions coupled with Mindlin's frictional analysis [5] also gives the same linear relationship [1] the interactions used in this model represent quite different physics on a much larger lengthscale. Typical values of δ_{max} observed in our simulations are $\sim 0.1 - 0.2\sigma$, which are several orders of magnitude larger than would be the case for a single point contact between elastic spheres of realistic elastic contacts (*e.g.*, for seeds or sand). In modelling surface displacements of single particle-particle contacts, a load independent contact stiffness is used *e.g.*, Hooke's Law constant, which arises from the material's elastic constants. In contrast, when modelling relative displacements of individual particles in an evolving assembly, the stiffness of the particle matrix is a strong function of the local voidage (*i.e.*, the assembly deformation) which in turn is related to the separation between particle centers. Hence, we require a displacement-dependent matrix stiffness, which is offered by the continuous-interaction model. The incorporation of assembly aspects into the pair interaction gives rise to effective bulk moduli which are much softer than that of the solid material, and typical maximum tangential displacements which are much larger [1].

To complete the interaction laws, we require the tangential stiffness k_T for the damping term in Equation (8), obtained by differentiating Equation (9), giving

$$C_T(\delta_F) = [mk_T(\delta_F)^{1/2}] = (\frac{3m\mu F_N}{2\delta_{max}}(1 - \frac{|\delta_F|}{\delta_{max}})^{1/2})^{1/2} \quad (11)$$

The particle configurations are generated by a time stepping algorithm [2].

ASSEMBLY DYNAMICS

We have carried out several DE simulation studies of slow shearing flows with this model [1] [2] [6] first with hoppers (filling and discharge) and then on heap formation by continuous vertical feed. In both cases the loads on the particles are comparitively low and the mechanism of assembly evolution is quite different to that taking place in compacted granular assemblies. In this case the granules execute highly cooperative movements through sustained quasi-static flow. (The exception is for the first few surface layers of heaps which are highly 'thermalised' and develop by a series of avalanches.) Many particles undergo a relative 'sliding' action with large bulk shear deformations whereas other regions of particles move *en masse* with relatively small relative displacements. These regions show up well in Figure 1 which gives sliding contact distributions of normalized contact tangential displacement vectors δ_F/δ_{max}, the ratio of tangential displacement to the maximum displacement required for the onset of gross sliding at that prevailing normal load. A funnel flow hopper is considered. This ratio can be used as a measure of how close a local region within the flow field is to rigid plastic failure or *rupture*. In Figure 1 we see that rupture zones, regions of high bulk shear deformation, are present in the CI ($n = 36$) case which allows for extended local connectivity, but are absent in a steeply repulsive Hertz interaction (Figure 1 (b)) which does not engage the local microstructure so well. There are areas in-between the high deformation regions in which the granules hardly move relative to one another and they move together essentially as a solid body. The CI interaction can be seen to generate regions of discontinuous assembly deformation, in keeping with experimental observation. In contrast the Hertzian particles are moving essentially independently, more like a liquid The regions of high strain for the CI case determine the overall flow field, and are the regions where particle degradation is most likely to occur and erratic flow to originate. Note that the rupturing is large near the walls.

Recent work on heaps has advanced our understanding of bed assembly mechanics. Figure 2 shows the density distribution of an evolving heap of monodisperse particles, indicated by center-to-center vectors. The heap is formed by freefall of model granules near to the center line [7]. The highly dilated region seen in the surface boundary layer is an avalanche. Avalanches do not form for much shorter range force interaction laws.

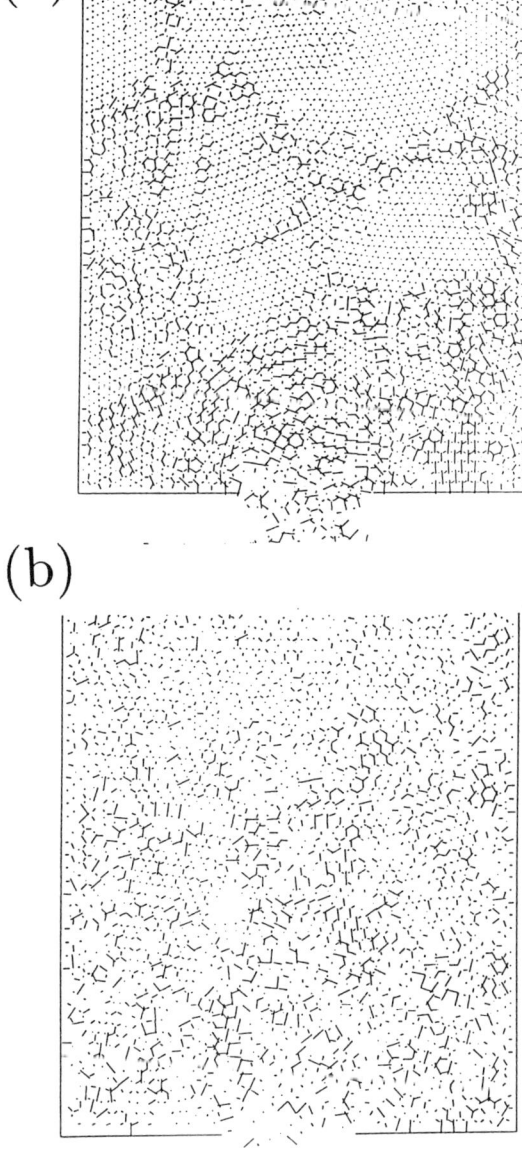

Figure 1 Tangential displacement ratio for a 2D Funnel Flow Hopper (a) with a Continuous Interaction, CI, interaction law and (b) with a stiff Hertz interaction law.

An avalanche is apparent as an open framework on the right of Figure 2, indicating that the density at the surfaces of the heap during an avalanche is much lower than at other places. The interior of the heap is largely unaffected during an avalanche; the avalanche is confined to a boundary layer near the surface. The effect of a series of avalanches is to smooth out local irregularities in the angle of repose on the heap surface. We found great difficulty in constructing heaps on a flat surface; stable heaps were found to form only on a geometrically rough base on the same scale as the particles. Heap formation is only weakly sensitive to the value of the contact friction coefficient. The appearance of avalanches, the pressure distribution on the base and the voidage distribution are sensitive to the analytic form of the elastic component of the normal interaction, with a soft-sphere r^{-36} potential giving more realistic behavior than an equivalent Hooke's law interaction with the same apparent spring constant [7].

The central region of the heap, around half-way between apex and base, has a low density (lighter shading) indicating a higher voidage with a patchwork of compacted regions formed close to the base. The packing density is highly non-uniform throughout the heap. In general, regions near the base of the heap are most densely packed, with looser packing towards the edges and apex. A resolution of the rotational motion of the particles, revealed that at the boundary between dilated and compacted regions, there is the most pronounced rotational activity with frequent counter rotation of adjacent particles in the boundary region being apparent. In fact rotational motion of the particles would appear to be an important mechanism by which the assembly relaxes, as it does not require appreciable center-of-mass motion. The comparative softness of the normal interaction ensures that each particle engages sufficiently strongly to most of its neighbors permitting significant transmission of stress through the assembly by rotational motion alone.

Another signature of large bulk deformation is the appearance of large variations in particle coordination number. This is illustrated in Figure 3 for a 2D heap, which shows the appearance of five co-ordinated particles in these regions.

Baxter *et al.*

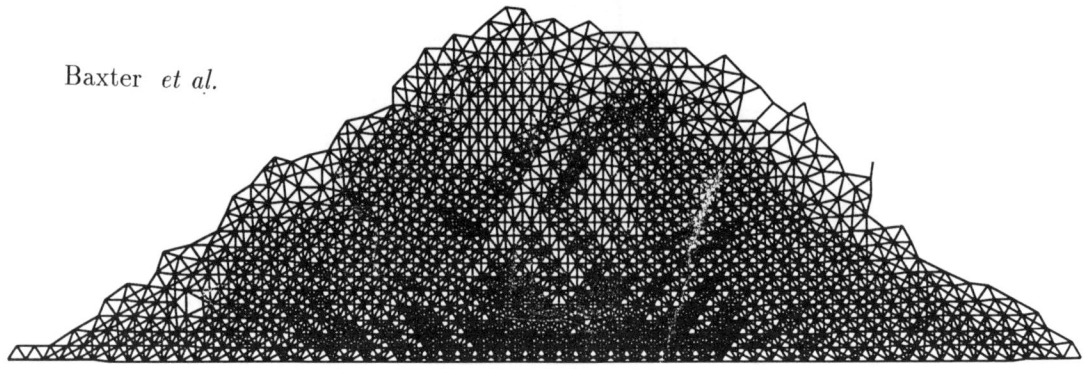

Figure 2 A 2D heap forming using the CI interaction law. Centre-to-centre 'bond' vectors between the particles are shown. Light regions indicate dilated material.

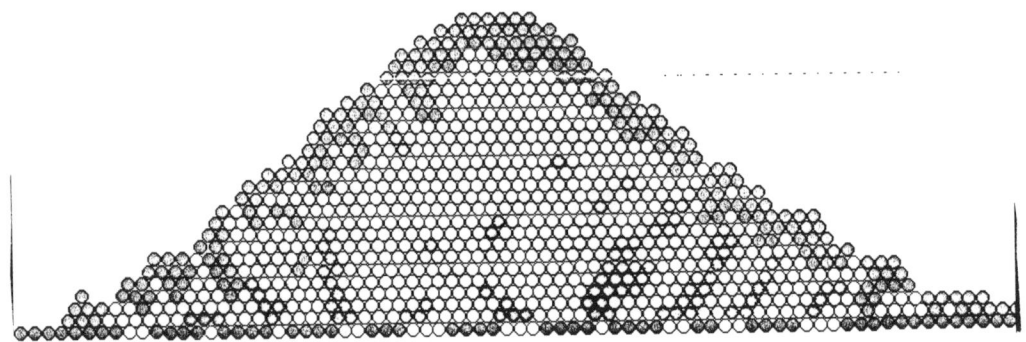

Figure 3 Co-ordination numbers of particles in a center-to-center vector plot for a Continuous Interaction simulation. Key: The darker disks on the outside of the heap indicate average coordination numbers less than 3. The darker disks in the central region of the heap indicate average coordination numbers ≥ 5. The light disks indicate a coordination number of 4.

LITERATURE CITED

[1] **Langston, P.A. Tüzün, U and Heyes, D.M.**, "Discrete Element Simulation of Granular Flow in 2D and 3D Hoppers: Dependence of Discharge Rate and Wall Stress on Particle Interactions," *Chem. Eng. Sci.* **50** pp. 967-987 (1995).

[2] **Langston, P.A. Tüzün, U and Heyes, D.M.**, "Discrete Element Simulation of Internal Stress and Flow Fields in Funnel Flow Hoppers," *Powder Technol.* **85** 153-169 (1995).

[3] **Walton, O.R.**, "Numerical Simulation of Inclined Chute Flow of Monodisperse Inelastic Frication Spheres," *Mech. of Mater.* **16** 239-249 (1993).

[4] **Johnson, K.L.**, "Contact Mechanics", 1st ed., p. 90, Cambridge University Press, Cambridge

[5] **Mindlin, R.D. and Deresiewicz, H.**, "Elastic Spheres in Contact Under varying Oblique Forces," *J. Appl. Mech. Trans. ASME* **20** 327-341 (1953).

[6] **Langston, P.A. Tüzün, U and Heyes, D.M.**, "Distinct Element Simulation of Interstitial Air Effects in Axially Symmetric Granular Flows in Hoppers", *Chem. Eng. Sci.* **51** 873-891 (1995).

[7] **Baxter, J. Tüzün, U Burnell, J and Heyes, D.M.**, "Granular Dynamics Simulations of Two-dimensional Heap Formation" *Physical Review E*, in press, (1997).

Electrostatic Effects in Cold-Model Circulating Fluidized Beds

Charles J. Coronella and Jianxun Deng

University of Nevada - Reno, Chemical & Metallurgical Engineering Department, Reno, NV 89557

The nature of particle/particle contact and particle/wall contact in the riser of a circulating fluidized bed (CFB) often leads to the presence of electrostatic potential in the beds. Although this triboelectric effect is well known to exist, few workers have investigated it further than simply making efforts to remove any electrostatics. Several methods are available for reducing electrostatics, including air humidification and the addition of small quantities of special conducting particles.

Despite these efforts, it is common for some residual electrostatic effects to linger in cold model test units, which are commonly made of nonconducting materials to enable visualization. This work focuses on the consequences of electrostatics on the hydrodynamics in the riser of a CFB. A dilute flow of sand is fluidized at a flux of 20 kg/(m^2s) in air flowing at a velocity of 3.8 m/s subject to various levels of humidity to control the degree of electrostatics. The effects on particle flux and pressure drop are presented.

As expected, increased electrostatics result in an increased pressure drop. The resulting electric field causes charged particles to be attracted to the column wall with opposite charge. This attraction has a more profound effect on smaller particles, leading to an apparent segregation of large and small particles in the riser. Near the riser wall, both the upward particle flux and the downward particle flux are increased significantly in the presence of significant electrostatic potential.

INTRODUCTION

In early studies on bubbling fluidized beds, some anomalous hydrodynamic phenomena were reported, which were often attributed to static electricity (Geldart, 1986; Osberg and Charlesworth, 1951; Rojo et al., 1986). The first reference on the static electrification of solids in fluidized beds appeared in the late 1940s to early 1950s. Since that time, a few papers on this topic have appeared (Briens et al., 1992; Guardiola et al., 1992; Baron et al., 1987; Guardiola et al., 1996). These authors found that the entrainment flux in the bubbling fluidized beds was greatly reduced when the relative humidity of the gas was increased. Some results indicated that electrostatic forces dominated the behavior of the smallest particles, and that as much as 50% of the measured pressure drop could be attributed to electrostatics. They concluded that static electricity probably accounts for the difficulty in predicting the flux of entrained particles from bubbling beds.

Compared with bubbling fluidized beds, the solids and gas in circulating fluidized beds have much higher velocities. The higher velocities cause substantially more particle-wall and particle-particle interactions, resulting in significantly increased electrostatic forces. Many laboratory-scale CFBs have risers made of nonconducting materials, e.g., Plexiglas, in order to visually observe the particles' complex hydrodynamic behavior.

University of Nevada, Reno

This allows for the build up of electrostatic charge to a substantial level. Other workers (Cheremisinoff, 1986; Hartge et al., 1985) have noted the existence of a strong static charge while measuring the voidage in the column using a capacitance probe or a photographic method. Chang and Louge (1992) concluded that it was almost impossible to carry out stable capacitance measurements when using plastic powder. They found that the pressure drop decreased by as much as 70% when they added into the column Larostat 519, an anti-static powder. This conclusion is consistent with that found in bubbling fluidized beds. Obviously, triboelectric phenomena play an important role in the hydrodynamics in many cold-model circulating fluidized beds.

In this work, some of the causes and effects of electrostatics are explored in dilute circulating fluidized beds. The static charge on the column surface was measured at different gas relative humidities. The effects of humidity on pressure profile, local solids flux, and particle-size distribution within the riser of a CFB are presented. This paper is merely an empirical report; an analysis of the complex triboelectric phenomena in gas-solid transport is given by Bailey (1984).

EXPERIMENTAL

The experimental apparatus is shown in Figure 1. The riser is made of Plexiglas, is 11.4 cm in diameter, and is 2.7 m tall. Gas-solid flow at the top of the riser is directed through a smooth glass elbow into a cyclone. Solids drop from the cyclone into a hopper that feeds a mechanical belt, which in turn feeds solids into the fluidized bed. Ten pressure probes are placed along the riser, and seven ports allowing for the measurement of the solids flux are available. A more detailed description of the system is available elsewhere (Deng, 1997).

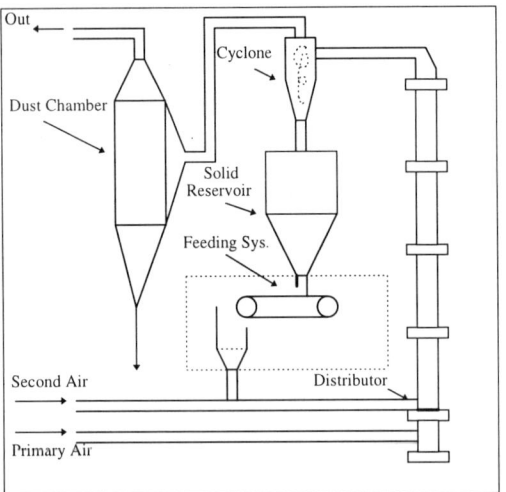

Figure 1 Schematic diagram of experimental apparatus

The mechanical belt system is designed to allow for direct control of the solids feed rate into the riser. Solids drop from the hopper onto the moving belt. An adjustable gate controls the height of the solids on the belt, and hence controls the rate of solids fed to the bed. Precise control of the feed rate is accomplished by varying the speed of the moving belt system. Solids fall from the belt through a coarse metal sieve, and finally drop into a horizontal air-driven pneumatic transport line, through which solids are fed into the base of the riser. The sieve is grounded and serves to remove static electricity.

All the experiments were carried out at ambient temperature and pressure and fluidized with air. The particles were silica sand with a size distribution given in Table 1, with a mean particle size of 209 μm. The particle density is 2599 kg/m^3.

The degree of static electrification was controlled by manipulating the humidity of the fluidizing air. Liquid water was pumped through a small water atomizer in the air stream. Not all the water that went into the air evaporated, but it was enough to control the relative humidity in the range of 7% - 14%.

Table 1

Size distribution of sand particles

size range	weight percent(%)
-500+355 μm	3.64
-355+212 μm	52.65
-212+180 μm	25.25
-180+63 μm	18.04
-63 μm	0.4

Particle flux within the riser was measured by two different types of probes. The first type, used to measure the flux of upward flowing particles, was made of a 6.5-mm inside-diameter tube 9 cm long in the axial direction. Sand and air passed into the probe, due to a positive pressure inside the riser. The contents of the probe flow into a small glass cyclone, where the solids are removed. The flow of gas leaving the cyclone was manipulated to ensure a nearly isokinetic measurement, as described by Herb *et al.*, (1992).

The second type of flux probe measured the flux of solids flowing downward, and was primarily used in the area near the riser wall. As shown in Figure 2, the probe was made from alternating sections of steel tubing and 400 mesh screen, all of diameter 8 mm. Particles that fall into the probe are captured into a reservoir to allow for subsequent weighing and screening. The alternating sections of the probe allowed for the gas flowing in the riser to pass by the probe, resulting in a flow pattern that is relatively undisturbed by the presence of the probe. Particles moving up will simply deflect off the sides of the probe, and particles moving down will fall into the circular end of the probe at the same flux as if the probe was not there.

A Chapman ESM500 electrostatic meter was used to measure the electric charge potential of the column wall. The range of the meter is -30 kV to + 30 kV, with an accuracy of ±5 %. The configuration of the measurements is shown in Figure 3. Pressure within the column was measured by a digital pressure gauge with 0.001 psi precision and accuracy of ±0.15 %. Relative humidity was measured by a TH550 digital gauge with accuracy of ±2%.

In all experiments reported here, the superficial gas velocity through the riser was 3.8 m/s, and the overall solids circulating rate was 20 kg/m^2 s. Before each experiment, the air and sand are circulated through the CFB

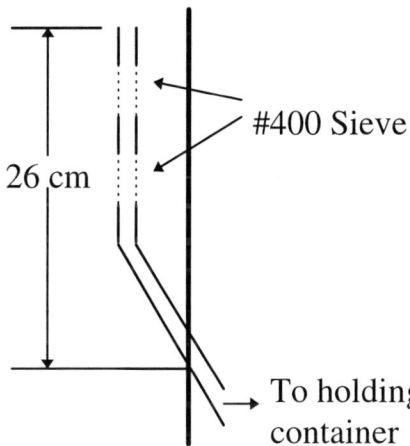

Figure 2 Schematic diagram of probe used to measure the flux of downward flowing particles.

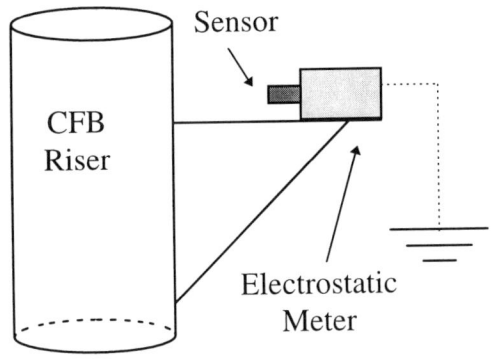

Figure 3 Schematic diagram of the electrostatic meter used to measure the electric charge of the column surface

for a minimum of 20 minutes in order to reach a steady state in the sand rate and in the air rate.

RESULTS

The static charge potential on the column surface at different axial locations and different air humidities is shown in Figure 4. These results show that the effect of increasing the humidity of the fluidizing air quickly results in decreased electrostatic potential, in agreement with Baron et al. (1987) and Guardiola et al., 1996. The electrostatic potential decreases with axial position. This is attributed to decreasing particle concentration with increasing axial position. A lower particle concentration results in fewer particle-particle collisions and fewer particle wall collisions. Each collision represents an op-

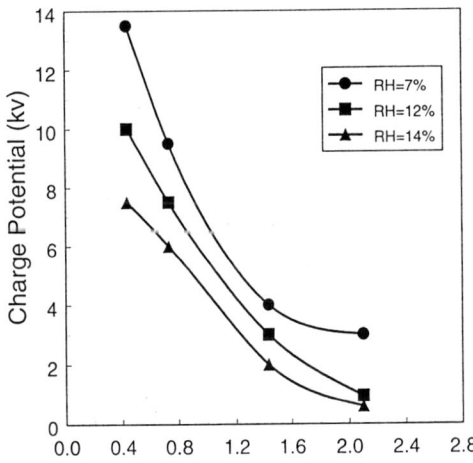

Figure 4 Static charge potential on the riser exterior surface as a function of humidity of the fluidizing air and the axial position.

portunity for electronic charge transfer, or increased electronic potential. An additional contributing effect is the particle velocity: near the base of the riser, the particle velocity has a large horizontal component, but that velocity quickly becomes vertical. Thus, particle-wall collisions are greatest nearest the base of the riser, resulting in increased opportunities for charge transfer.

Baron et al. (1987) showed that electrostatic forces contribute significantly to the pressure drop through the freeboard of a bubbling fluidized bed, and Ally and Klinzing (1985) reported that electrostatic charging, controlled by humidity, has a profound effect on the pressure drop in dilute-phase pneumatic transport. Chang and Louge (1992) noticed that the electrostatic phenomena can greatly affect the pressure drop through the riser of a CFB. Figure 5 shows the pressure gradient profile in the riser at three humidities. Obviously, the effect of increased humidity (and decreased electrostatic potential) is to decrease the pressure gradient. This effect is most notable near the base of the column, due to the reasons discussed above. It is interesting to note just how quickly a small change in humidity causes a change in the pressure gradient. As sand moves through the riser, it "takes" electrons from the column wall, and so becomes negatively charged, while the inside surface of the wall becomes positively charged (Bailey, 1984). The resulting attractive force causes the velocity of the sand particles to decrease in the riser, resulting in a greater particle concentration. This increased particle concentration is what causes the increase in pressure gradient seen in Figure 5.

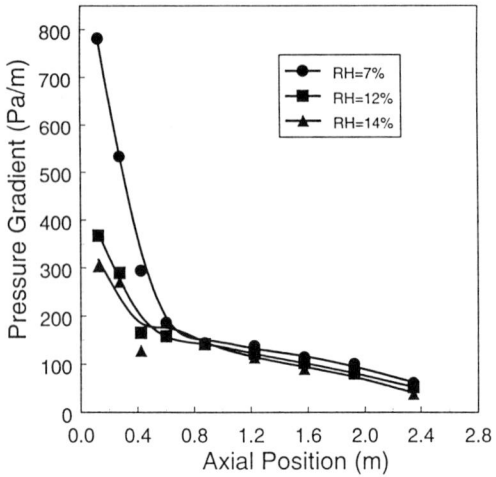

Figure 5 Pressure gradient as a function of relative humidity and axial position.

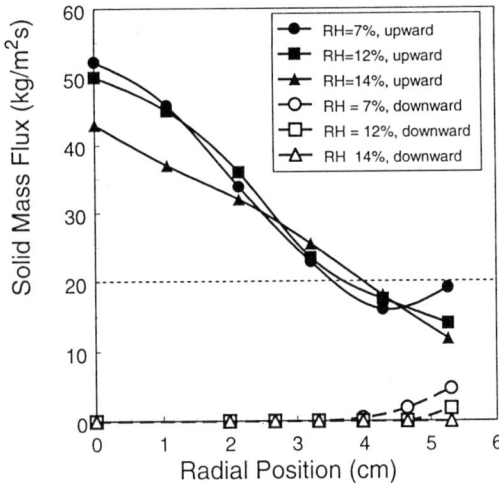

Figure 7 Upward and downward particle flux profiles at an axial position of z = 1.66 m, as a function of relative humidity and radial position.

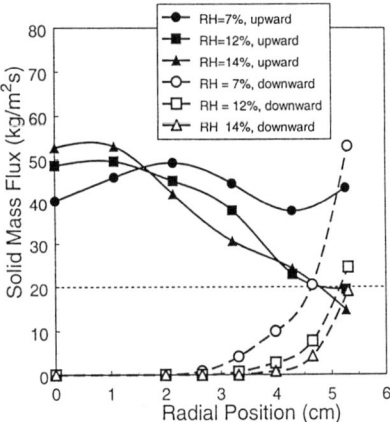

Figure 6 Upward and downward particle flux profiles at an axial position of z = 0.66 m, as a function of relative humidity and radial position.

Figures 6 and 7 show the radial flux profiles of particles flowing at axial positions of z = 0.66 m and z = 1.66 m, respectively, each at three humidities. Near the base of the column, Figure 6 shows the upward flux profile is nearly flat when the relative humidity is 7%, corresponding to the case of highest electrostatic potential. As the humidity increases, the upward flux gradually changes to the more typical profile, with a maximum flux near the center of the column, and a minimum flux near the column wall. The downward flux profiles show very little downward movement in the column, except near the riser wall. The downward flux is much greater in magnitude near the base of the column (Figure 6) then it is further up (Figure 7).

In Figure 6, the upward flux for RH = 7% has a greater upward flux compared to the other humidities everywhere except near the center of the riser cross section. Consequently, the total upward solid flux, proportional to $\int_0^R (G_p\, r)\, dr$, is greatest for this case. Figure 6 also shows the downward flux profiles, where it is apparent that the downward flux is greatest in this case. By continuity of mass, the average *net* flux (defined as the difference between the upward flux and the downward flux) at any axial position must be constant and equal to the overall solids circulating rate of 20 kg/(m^2 s). Therefore, the downward flux must be maximum for the case of RH = 7%.

At the lower axial position, the effect of humidity (i.e., decreased electrostatic potential) is to decrease both the upward and downward solids fluxes. Particle flux is proportional both to particle concentration and to particle velocity. It seems unlikely that a decrease in humidity or an increased electric field could cause an increase in particle velocity. Therefore, the measurements imply that the solids concentration is increased greatly in the case of significant electrostatic effects, especially in the region near the wall. It is interesting to note that Figure 7 shows little dependence of the solids flux on humidity at a higher axial position. This may be attributed to the reduced particle concentration at a higher axial position, as discussed above.

Figure 8 shows the net particle flux profiles. The downward flux profiles were interpolated by a cubic spline to the radial positions where the upward flux was

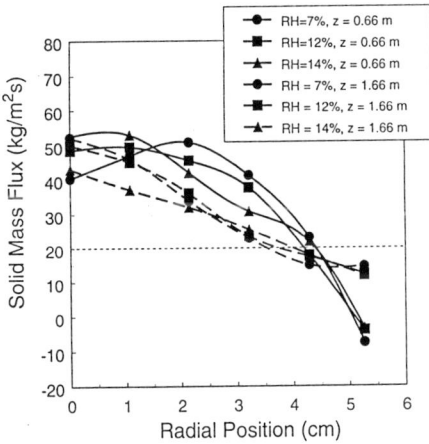

Figure 8 Net particle flux profiles at z = 0.66 m and at z = 1.66 m and as a function of humidity.

measured, allowing for calculation of the downward flux by subtraction. Apparently, the humidity has little effect on the net flux. Increased electrostatic potential causes increased upward and downward fluxes in the same locations. This is further confirmation of the conclusion that the increased electric field causes an increase in particle concentration near the column wall.

Table 2 shows the size distribution of particles collected in the upward flux probe at the two axial positions for two different cases of humidity near the column wall (r = 5.3 cm). At both axial positions, the fraction of fine particles collected from the probe was increased when the humidity was decreased. An obvious conclusions is that the smallest particles are most strongly affected by the electrostatic forces. Note that in all four cases shown, the fraction of large particles is larger than the fraction given in Table 1 for the size distribution of the sand used in these studies. This is indicative of the high rate of internal recirculation of the largest particles in a CFB riser, and is the subject of other publications (Deng, 1997; Coronella and Deng, 1997).

CONCLUSIONS

Some of the effects of humidity and triboelectric phenomena were investigated in a dilute CFB. The degree of effects of electrostatics within a CFB riser may be controlled by manipulating the humidity of the fluidizing air. Electrostatic phenomena are most significant where the particle concentration is greatest, near the base of the column. With a low relative humidity, the pressure gradient increases significantly. Using a particle flux probe, it was shown that both the upward and downward flux increase near the riser wall in the absence of humidity. The net flux, however, is barely affected by the electrostatic phenomena. An examination of the size distribution of particles recovered from the probe that measures upward particle flux shows that the concentration of fine particles is greatly increased near the wall in the presence of strong electrostatic potential.

Table 2

Particle size distribution of sand particles captured in the upward flowing flux probe at the column wall (r = 5.3 cm).

Axial Position (m)	RH (%)	+355 µm (wt%)	-355+212 (wt%)	-212+180 (wt%)	-180 µm (wt%)
0.66	7	4.24	58.99	20.68	16.09
0.66	14	6.04	62.97	21.32	9.67
1.66	7	4.50	59.13	17.75	18.59
1.66	14	5.78	59.95	21.02	13.25

SYMBOLS

G_p Local flux of particles (kg/m^2 • s)
R   Radius of the riser (m)
r   Radial position (m)
RH   Relative humidity (%)
z   Axial position (m)

ACKNOWLEDGMENT

Grateful acknowledgment is made for the financial support provided by the Graduate School at the University of Nevada, Reno.

REFERENCES

Ally, M. R., and Klinzing, G. E., 1985 "Inter-relation of Electrostatic Charging and Pressure Drops in Pneumatic Transport" *Powder Technol.* **44** p. 85.

Bailey, A. G., 1984 "Electrostatic Phenomena During Powder Handling" *Powder Technol.* **37** p. 71.

Baron, T., Briens, C. L., Bergougnou, M. A., and Hazlett, J. D., 1987 "Electrostatic Effects on Entrainment from a Fluidized Bed" *Powder Technol.* **57** p. 55.

Briens, C. L., Bergougnou, M. A., Inculet, I. I., Baron, T., and Hazlett, J. D., 1992 "Size distribution of particles entrained from fluidized beds: electrostatic effects" *Powder Technol.* **70** p. 57.

Chang, H., and Louge, M., 1992 "Fluid dynamic similarity of circulating fluidized beds" *Powder Technol.* **70** p. 259.

Cheremisinoff, N. P., 1986 "Review of Experimental Methods for Studying the Hydrodynamics of Gas-Solid Fluidized Beds" *Ind. Eng. Chem. Proc. Des. Dev.* **25** p. 331.

Coronella, C. J. and Deng, J., 1997 "Particle Segregation in Dilute-Phase Circulating Fluidized Beds" in preparation.

Deng, J., 1997 *Hydrodynamics of a Cold Circulating Fluidized Bed* Ph. D. Diss., University of Nevada, Reno.

Geldart, D., 1986 *Gas Fluidization Technology* Wiley, New York.

Guardiola, J., Ramos, G., and Romero, A., 1992 "Electrostatic Behavior in Binary Dielectric/Conductor Fluidized Beds" *Powder Technol.* **73** p. 11.

Guardiola J., Rojo, V., and Ramos, G., 1996 "Influence of Particle size, fluidization velocity and relative humidity on fluidized bed electrostatics" *J. Electrostatics* **37** p. 1.

Hartge, E. U., Li, Y., and Werther, J., 1985 "Analysis of the Local Structure of the Two Phase Flow in a Fast Fluidized Bed", *Circulating Fluidized Bed Technology II*.

Herb, B., Dou, S., Tuzla, K., and Chen, J. C., 1992 "Solid mass fluxes in circulating fluidized beds" *Powder Technol.* **70** p. 197.

Osberg, G. L. and Charlesworth, D. H., 1951 "Elutriation in a Fluidized Bed", *Chem. Eng. Prog.* **47**.

Rojo, V., Guardiola, J., and Vian, A., 1986 "A Capacitor model to interpret the electric behaviour of fluidized beds. Influence of apparatus geometry" *Chem. Eng. Sci.* **41**(8) p. 2171.

A New Computer Controlled Wurster-type Particle Coating Apparatus

H. Littman
Department of Chemical Engineering, Rensselaer Polytechnic Institute, Troy, NY 12180

M. H. Morgan III,
Department of Chemical Engineering, Hampton University, Hampton, VA 23668

C. B. Morgan
Department of Mathematics, Hampton Universtiy, Hampton, VA 23668

This paper describes a new Wurster-type particle coating apparatus suitable for coating both coarse and fine particles. It features a dual air jet and coating spray arrangement, and employs a PC computer to automatically monitor and control all essential parameters of the coating operation. The coating cannot be damaged by wall collisions as the particles are recirculated in the apparatus and particle elutriation has been eliminated without the use of cyclones or filters. To test the apparatus, raw crystalline aspirin powder with an average diameter of 186 μm was coated with a solution containing Eudragit L30D.

A Wurster-type coater is fluid mechanically a draft tube spout-fluid bed (DTSFB). The relevant characteristics of a DTSFB which account for its flexibility in meeting process coating conditions are discussed to provide a clear understanding of the important features of the coater. Drying in the spout and annulus are discussed and a mathematical model presented to predict the particle coating thickness distribution.

This paper describes a new Wurster-type particle coating apparatus (1) suitable for spray coating both coarse and fine particles. Since the coating of fine particles is particularly challenging, this paper will describe solutions to problems associated with the coating of raw crystalline aspirin powder with an average particle diameter of 186 μm. A mathematical model gives the mean and variance of the particle coating thickness distribution. Conventional coaters are improved models of the original design patented by Wurster (2).

The coater (Fig. 1) features a dual air jet and coating spray arrangement, and employs a PC computer to automatically monitor and control all essential parameters of the coating operation. There is a completely integrated computer control system[*] that provides for automatic and continuous monitoring of all fluid flowrates, temperatures and critical pressure measurements throughout the system. These outputs are electronically monitored and the resulting data used to control the coating process by making real time adjustments to critical input parameters. Important dimensions of the experimental coater are listed in Table 1.

There are two coating zones, and two heating and drying zones. The particle circulation rate, and the fountain height of particles leaving the draft tube are both controlled. Nowhere in the apparatus do particles collide at high velocities with solid surfaces as they recirculate, and particle elutriation from the coater is eliminated without the use of cyclones or filters.

Particles, screened to obtain a narrow size range, are spray coated at the entrance to the draft tube and surface dry as they pass upward. With fine particles, the solids leaving the draft tube are redirected into the annulus by a downward flow of air in the fountain tube and dry completely in the annulus. The particles reenter the draft tube from the bottom of the annulus. Note in Fig. 1 that the particles leaving the draft tube can undergo additional drying and/or coating above the draft tube giving the unit a high drying capacity. Drying in the moving packed bed annulus is particularly effective because of the long residence time and the high heat and mass transfer coefficients which occur there.

The temperature of the hot air streams, $V_A(0)$ and $V_N(0)$ in Fig. 1, determine the air and particle temperatures entering the draft tube. As a coated particle rises in the draft tube, water evaporates from its surface due to the difference in water partial pressure between the particle surface and the bulk air phase. The heat required for evaporation initially comes from the bulk air phase but as the particle surface cools, heat flows from the interior of the particle. Thus, the particle surface temperature falls with distance from the inlet to the draft tube but remains above the wet bulb temperature.[**] Once the particle surface becomes dry, the moisture must diffuse through coating and this slower process is completed in the annulus. The coating solution feed rate is not limited by the requirement that the particle be surface dry before it leaves the draft tube because drying can occur in the fountain tube region.

RELEVANT CHACTERISTICS OF THE DRAFT TUBE SPOUT-FLUID BED (DTSFB) FOR COATING APPLICATIONS

From a fluid-particle systems standpoint, a Wurster coater is a draft tube spout-fluid bed (DTSFB). The attractiveness of this system for coating applications lies in the ease with which the particle circulation can be controlled and its flexibility in meeting process specifications. All of the improvements mentioned in this paper over conventional Wurster coaters result from

---

[*] Page limitations do not permit details of the control system or detailed modeling of the heat and mass transfer in the draft tube and annulus.

[**] Markowski (3) shows that this phenomenon occured in his jet spouted bed solution drier.

exploitation of the fluid mechanical characteristics of the DTSFB. It is for this reason that much of the paper is devoted to the mechanics of the DTSFB.

In the classical spouted bed (SB), a jet of fluid penetrates a bed of particles creating a central spout zone and an annulus region surrounding the spout. The particles are transported upward through the spout and then recirculate down the annulus. Particles feed into the spout at the spout-annulus interface as gas passes in the opposite direction from the spout into the annulus. If the total gas flow at inlet is divided between the spout and annulus regions, the bed operates as just described but is called a spout-fluid bed (SFB). Flow regimes for SB's and SFB's can be found in the literature (4,5).

Although SB's are commonly thought of as being useful only for processes involving coarse particles (approximately 1 mm and larger) smaller particles can be spouted using a draft tube which acts as an artificial spout. The particles have a short residence time in the draft tube (and fountain tube) where they must surface dry to prevent their sticking together when they fall to the top of the annulus. The residence time in the annulus is much longer which allows the coating to dry completely before recoating occurs.

The draft tube does away with the maximum spoutable height limitation of SB's and SFB's so that there is no limit on the annulus bed height. In addition, the solids flowrate is decoupled from the gas flowrate allowing control over the both flowrates in the draft tube. This is accomplished by regulating the pressure at the inlet to the draft tube through variation of l and $V_A(O)$ in Fig. 2. Two additional benefits of the draft tube are the narrowing of the residence time distributions of gas and particles and a significant reduction in the system pressure drop over that in a spouted bed.

When the jet inlet tube to particle diameter ratio exceeds about 20, bubbling instabilities are introduced into the spout of a SB. These instabilities will occur closer and closer to the spout inlet tube as the particle size is reduced and/or the annulus flowrate is increased. This instability can be overcome by increasing $V_N(O)$. The relevance here is that the entry region to the DTSFB is basically as a shallow spouted bed. A detailed discussion of the spouting of fine particles is given by Cecen et al (6).

Coarse particles of the order of 1 mm entrain smoothly into the draft tube at gas velocities greater than terminal without flow instabilities. When the particle size is reduced, the aforementioned instabilities in the form of bubbles originate between the outlet of the jet inlet tube and the inlet to the draft tube. These bubbles carry particles in their wake which transport up the draft tube if the gas velocity is high enough. An example of this is the slugging draft tube coal gasifier described by Judd and Rudolph (7). The frequency at which the bubbles are generated becomes the frequency of the pressure and voidage fluctuations in the dense phase near the inlet to the draft tube. This flow condition is detrimental to producing a uniform coating on each particle and increases particle agglomeration in the draft tube. To avoid this with the fine aspirin particles, the jet flow [$V_N(0)$] was increased to 5 to 10 m/s which is 6 to 10 times above the terminal velocity of the aspirin. At these air velocities, pressure and voidage fluctuations disappear in the draft tube and the particles transport up the draft tube in dilute phase flow at high voidages. These high velocities with fine particles would result in a high fountain height in the absence of the downward fountain flow. The fountain flow not only limits the fountain height and but directs the particles on to the top of the annulus without high velocity impacts on any wall or filter.

OTHER APPLICATIONS OF DTSFB'S

Buchanan and Wilson (8) called the DTSB a fluid-solids lift recirculator and showed that it was more effective than a SB with respect to its circulation efficiency. Schwedes and Richter (9) and Krambrock (10) used this system as a particle blender and there are many units of this type in commercial use for that purpose. Hattori and Takeda (11) investigated the side outlet spouted bed with a draft tube and showed that it has excellent potential as a chemical reactor. Claflin and Fane (12,13) have used a draft tube system for drying and the thermal treatment of grains. Khoe and Van Brakel (14) investigated the draft tube spouted bed as a small scale grain dryer.

Grbavcic et al. (15) have compared the basic fluid mechanical characteristics of SB's and SFB's with the characteristics of draft tube systems and concluded that the recirculation efficiency is up to 5 times higher than that in a SB. In addition, they showed that the DTSFB is a very flexible fluid-particle contacting system because the fluid flowrates through the draft tube and annulus, and the solids circulation rate can be very easily controlled and adjusted in accordance with process needs.

Brereton et al. (16) developed an axisymmetric model of a draft tube system by assuming a constant pressure boundary condition in the inlet section. Berruti et al. (17) investigated solids circulation in air-sand SB's and SFB's with a draft tube and concluded that the solids circulation rate is strongly affected by the geometry of the system and by the inlet spouting and annulus flowrates. Matthew et al. (18) investigated a liquid phase spouted bed with a draft tube and described the hydrodynamics in the inlet section using an axisymmetric model which takes into account non-Darcy flow. They found good agreement of their experimental pressure and streamfunction data with their theoretical predictions, and also concluded a) that the inlet section behaves as a classical SB and b) that the Morgan-Littman boundary condition (19) can be used to describe the pressure variation along the spout-annulus interface. A similar hydrodynamic model for flow in the annulus of a SFB is given by Day et al. (20). The DTSFB system investigated by Yang and Keairns (21) differs from the normal DTSFB in that their annulus is fluidized and the draft tube totally immersed in the bed.

The DTSFB used by Hadzismajlovic et al. (22) for drying dilute aqueous solutions containing yeast has characteristics so relevant to this coater that will be discussed in some detail. Possible uses of the solution drier mentioned in the paper include the drying of yeast, whole or skim milk, corpuscles, whole blood, tomato and citrus juices, egg albumen, soy milk and tea. The drier was found to have a higher evaporative capacity than a SB drier and a higher volumetric capacity than a spray drier. The quality of the solids was about the same as that from a spray drier.

The solution to be dried was fed in at the top of the annulus so that most of the drying occurred in the annulus. Hot air streams, $V_N(0)$ and $V_A(0)$, were introduced through a nozzle at the bottom of the drier and into the annulus through a perforated cone. These flowrates and their temperatures were independently controlled. Solution containing yeast solids at about room temperature were sprayed on to the top of the annulus and then carried downward as a film coating polypropylene particles [3.6 mm equivalent sphere diameter]. The particles circulated from the annulus into the entry section and then through the draft tube. The dried yeast in the solution deposited on the particles and was subsequently flaked off when the particles struck a deflector plate located above the draft tube. The yeast was then elutriated from the drier and collected using a cyclone or bag filter. The uncoated particles then returned to the top of the annulus to be recoated.

For the three highest liquid flowrates used, the nozzle air, $V_N(0)$, entering the bed cooled rapidly in the inlet section and in the lower part of the draft tube due to the evaporation of water from the thin liquid film coating each particle. The water was completely evaporated somewhere in the draft tube after which the particle began to heat the surrounding air. Farther up the draft tube, the air and particle temperatures equilibrated. For the lowest liquid flowrate used, there was no minimum in the axial gas temperature profile as the particles dried completely in the annulus.

HYDRODYNAMIC MODELING

The hydrodynamic modeling of the DTSFB has been addressed for gas and liquid phase systems involving large particles (23,24,25,26). First considerations involve the prediction of solids transport in a draft tube using one dimensional models. Expressions for the drag coefficient and wall friction are given which complete the modeling for large spherical particles. The region below the inlet to the draft tube where the spout and annulus flowrates [$V_A(O)$ and $V_N(O)$] are redistributed to provide the fluid velocity in the draft tube, U_d, and in the annulus surrounding the draft tube, U_a, has also been addressed (24). This redistribution is not, in general, predictable but must be obtained experimentally using a measurement of the annulus pressure drop.

An important aspect of the hydrodynamics which accounts for superior performance of the DTSFB coater is the ease with which the particle mass flowrate through the draft tube can be controlled. The annulus air stream, $V_A(0)$, allows the gas velocity to vary between 0 and U_{mF} in the annulus. Any change that increases the pressure at the inlet to the draft tube increases particle circulation rate. Normally that is accomplished by increasing $V_A(0)$ and l. Grbavcic et al. (24) showed in a liquid-solid DTSFB that the solids mass flowrate increased rapidly with l for a fixed value of $V(0)$ and $V_A(0)$. In such a case U_a increases as the height of the spout-fluid bed inlet section is increased. This higher velocity raises the annulus pressure drop causing the pressure at the inlet to the draft tube to rise. Continuity, however, requires that the fluid flowrate in the draft tube diminish. Model calculations given in Grbavcic et al. (24) show that the rise in solids fraction is more important than the decrease in solids phase velocity so that the mass flowrate rises. Berruti et al. (17) found similar results for air-sand spouted and spout-fluid beds.

When $V_A(0)$ is raised keeping $V(0)$ and l constant, similar results are obtained to that described above. The effect is even more pronounced when $V_N(0)$ and l are constant and $V_A(0)$ is raised. In Hadzismajlovic et al (22), the solids mass flux changed by a factor of about 4 when $V_A(0)$ was raised from zero to that which causes the annulus to fluidize. An important but not obvious result found by Grbavcic et al. (24) occurs when $V(0)$ is increased keeping $V_A(0)$ and l constant. In that case U_a decreases and with it the pressure drop across the draft tube and annulus. Nonetheless the particle mass flowrate in the draft tube increased because an increase in particle phase velocity [due to an increase in $V_N(0)$] outweighed a solid fraction decrease. Thus we see that there are a variety of ways in which the particle circulation rate can be varied in a DTSFB. Gas phase and liquid phase systems do not behave identically as the drag coefficient and wall friction forces differ. These effects will vary widely with particle size and density in gas systems and must be dealt with experimentally at present. The state of development of the DTSFB contactor for large particles is given by Littman (28).

Another effect of particle size on the design is worth mentioning. Since the inlet air velocity must be of the order of the terminal velocity and the maximum annulus velocity is the minimum fluidizing velocity reducing the particle size will change the ratio of U_t to U_{mF}. As U_t/U_{mF} increases, the annulus cross sectional area relative to that of the draft tube must increase to balance evaporation in the annulus and spout.

Finally, it is important to recognize that the higher the inlet air temperature of $V_N(0)$ the greater the heat input to the drier and, thereby, its evaporative capacity. High air temperatures can be introduced into a DTSFB without damaging heat sensitive particles provided that those particles are not dry when in contact with the hottest air.

ADVANTAGES OF THE NEW WURSTER-TYPE COATER

1. There is independent control of flowrate and temperature of all feed streams.
2. There are two high relative velocity spray coating zones within the unit.
3. There are two heating zones for evaporating the coating liquid. Within each zone the air temperature can be varied and controlled.
4. Particle elutriation from the bed is eliminated without the use of cyclones or filters. This is made possible by varying the diameter of the vent header and by using a narrow particle size range.
5. There is a completely integrated computer control system that provides real time monitoring of all fluid flowrates, temperatures and critical pressure measurements throughout the unit. These outputs are used to control the coating process by making real time adjustments to critical input parameters. The basis for control lies in the ease with which the pressure at the inlet to the draft tube can be varied.
6. The fountain tube serves as an additional spray coating zone and low impact particle separator.

This design can be compared with the industrially significant Glatt Air Techniques coater. In that design, air is drawn through a distributor plate with a fixed hole pattern. A spray nozzle is fixed in the center of the distributor plate and the draft tube located above the plate. A filter above the exit to the draft tube prevents particles from leaving the DTSFB section. In contrast with the new Wurster design, the Glatt coater a. lacks flexibility because the draft tube and annulus flowrates cannot be independently varied, b. the gas and particle flowrates are coupled, c. with fine particles coating is ineffective and a filter is needed to limit the fountain height, d. there is no comparable data acquisition and control system and, e. there is only a single heating and spraying zone.

OPERATIONAL PROBLEMS THAT WERE OVERCOME TO COAT THE ASPIRIN PARTICLES

1. The coater was charged with 2.5 kg of raw crystalline aspirin powder screened to have a narrow size distribution and an average diameter of 186 µm. The raw material was rodlike with a length of diameter ratio of about 2 to 1. The particle density was 1390 kg/m^3. While a wide size distribution is usually undesirable from a product standpoint, such a distribution is particularly undesirable in the coater because elutriation of fines

from the bed must be controlled. In the new design, the gas velocity in the vent header is maintained below the terminal velocity of the finest particle.

2. The coating solution contained 50.7% Eudragit L30D, 3.04% PEG 8000, 3.04% micronized talc, 0.16% Keltrol F and 43.1% purified water. It flowed well through the air atomizing nozzle used in this work (Spraying Systems - 1/8 JJ nozzle assembly J11-SS with a J2050 fluid cap and J67147 air cap). A peristaltic pump was used to pump the solution through the fluid cap where it emerged as a liquid jet. The jet was atomized by a turbulent airstream flowing cocurrent to it. When plugging of the nozzle was encountered after 10 to 15 minutes of operation, the diameter of the hole in the fluid cap was increased from 0.51 to 1.52 mm and the air cap hole machined to a conical shape. With the redesigned nozzle, no plugging occurred in two hours of operation. A rough estimate of the droplet size indicated that 20% of the droplets were finer than 100 mm and 80% smaller than 200 mm. The droplet size is adjustable by varying the air atomizing pressure and coating solution flowrate.

3. After observing pressure fluctuations in the inlet to the draft tube, the air flowrate was increased from the range of 10 to 30 scfm to 50 to 100 scfm. At these flowrates, the fountain height is extremely large requiring an increase in the fountain flowrate $V_F(0)$ from the range of 10 to 20 scfm to 30 to 60 scfm to maintain it at its original height.

MATHEMATICAL MODEL OF THE COATING THICKNESS DISTRIBUTION

A stochastic model based on renewal theory concepts was presented by Morgan and Morgan (29,30) to simulate the real time growth of the coating thickness on particles in batch units. They determined that a gamma distribution represented the cycle time distribution (CTD) in a DTSFB coater. The probability density function is

$$f(t) = \frac{\lambda(\lambda t)^{\gamma-1} e^{-\lambda t}}{(\gamma-1)!} \quad (1)$$

where λ, γ and t are all greater than zero. The data of Mann and coworkers (31,32,33) were used to show that $\gamma = 6$ and the excellent fit of the CTD data for three different liquid flowrates is seen in Fig. 3. Based on the method of maximum likelihood (34), λ is related to the mean and variance of the gamma distribution by

$$\lambda = \frac{\mu_t}{\sigma_t^2} \quad (2)$$

Using specified bed geometry and particle parameters for the DTSFB, Morgan and Morgan (29) were able to predict μ_t and σ_t^2 completing the specification of all the parameters in eqn 1. t_{max} and t_{min} in Fig. 3 were obtained from Mann and Crosby's data but they can be calculated from the basic mechanics of the particle flow in the annulus.

The particle coating distribution (PCD) and the cycle time distribution (CTD) are related by the following equations (30)

$$\mu_c(t) = \frac{\mu_w}{[\mu_t/t]} \quad (3)$$

$$\sigma_c^2(t) = \frac{\sigma_w^2}{[\mu_t/t]} + \frac{\mu_w^2 [\sigma_t^2/t^2]}{[\mu_t/t]^3} \quad (4)$$

μ_w and σ_w^2 are the mean and variance of the equivalent coating thickness deposited on a particle in a single passage through the draft tube. These quantities are associated with the characteristics of the spray nozzle and its location in the apparatus.

Morgan and Morgan (30) then investigated the use of their gamma distribution for the CTD to simulate the PCD for a range of operating times and bed sizes and compared their results with Mann and Crosby's predictions. Those investigators had shown that for times 50 times greater than the mean (μ_t), the PCD asymptotically approaches a normal distribution. The simulation agrees with the Mann and Crosby's prediction even when t is small. The simulations show that only μ_w is important in determining the PCD.

SUMMARY

1. A new computer controlled coating apparatus of the Wurster-type is described for coating both coarse and fine particles.
2. The relevant mechanics upon which the coater is based is discussed in detail.
3. A mathematical model is presented for predicting the mean and variance of the particle coating thickness.

EQUATIONS AND SYMBOLS

CTD	-	cycle time distribution
d_f	-	diameter of fountain tube, m
d_i	-	diameter of bed inlet tube, m
D_c	-	annulus diameter, m
D_d	-	diameter of draft tube, m
$f(t)$	-	probability density function for CTD
H_a	-	height of the annulus, m
H_f	-	fountain height, m
l	-	height of the entry section of the DTSFB, m
l_d	-	length of the draft tube, m
L_f	-	length of fountain tube, m
L_i	-	length of bed inlet tube, m
PCD	-	particle coating distribution
t	-	time
t_{min}	-	min. particle circulation time, s
t_{max}	-	max. particle circulation time, s
U_a	-	superficial fluid velocity in annulus, m/s
U_d	-	superficial fluid velocity in draft tube, m/s
U_{mF}	-	minimum fluidizing velocity, m/s
U_t	-	terminal velocity, m/s
V(0)	-	total volumetric flowrate at inlet of DTSFB, $V_A(0)+V_N(0)$, m^3/s
$V_A(0)$	-	volumetric fluid flow to annulus at inlet to DTSFB, m^3/s
$V_F(0)$	-	volumetric fluid flow in fountain tube, m^3/s
$V_N(0)$	-	volumetric fluid flow to spout at inlet to DTSFB, m^3/s
W_d	-	solids mass flux in draft tube, kg/m^2,s

Greek letters

γ	-	parameter in eqn. (1)
λ	-	parameter defined by eqn. (2)

μ_c	-	mean of PCD, m
μ_t	-	mean of PTD, s
μ_w	-	mean equivalent coating thickness, m
σ_c^2	-	variance of PTD, m^2
σ_t	-	variance of CTD, s
σ_w^2	-	variance of equivalent coating thickness, m^2

Fig. 1

AF	-	air filter
MFC	-	mass flow controller
PP	-	pressure probe
PR	-	pressure regulator
TC	-	thermocouple probe

LITERATURE CITED

1. **Littman, H., M.H. Morgan III and S.Dj. Jovanovic**, US Patent 5254168, 1993.

2. **Wurster, D.E.**, US Patent 3089824, 1963.

3. **Markowski, A. S.**, "Drying Characteristics of a Jet Spouted Bed Drier", Can. J. Chem. Eng. (1992) 70, 932-944.

4. **Littman, H., R. Zamora and M.H. Morgan III**, "The Spouting Regime Map, Bubbling Characteristics and Bed Expansion in Deep Beds of Large Hollow Polyethylene Spheres", Can. J. Chem. Eng. (1989) 67, 912-915.

5. **Hadzismajlovic, Dz.E., D.V. Vukovic, Z.B. Grbavcic, R.V. Garic and H. Littman**, "Flow Regimes for Spout-Fluid Beds", Can. J. Chem. Eng. (1984) 62, 825-829.

6. **Cecen, A., H. Littman and M.H. Morgan III**, "Flow Regime Diagrams and Stability of Fine Glass Spheres Spouted with Air", Fluidization VIII, Eng. Found., N.Y. (1996) 207-216.

7. **Judd, M. R. and V. Rudolph**, "Gasification of Coal in a Fluidized Bed with a Draft Tube", Fluidization V, Eng. Found. N.Y. (1986) 505-512.

8. **Buchanan, R.H. and B. Wilson**, "The Fluid-Solids Lift Recirculator", Mech. Chem. Eng. Trans. (Australia) (1965) 1, 117-124.

9. **Schwedes, J. and W. Richter**, "Der pneumatische Granulatmischer", Aufbereitungs-Technic (1976) 17, 115-119.

10. **Krambrock, W.**, "Mixing and Homogenizing of Granular Bulk Material in Pneumatic Mixer Unit", Powder Technol. (1976) 15, 199-206.

11. **Hattori, H. and K. Takeda**, "Side-Outlet Spouted Bed with Inner Draft-Tube for Small-Sized Solids Particles", J. Chem. Eng. Japan (1978) 11, 125-129.

12. **Claflin, J.K. and A.G. Fane**, "The Use of Spouted Bed for the Heat Treatment of Grains", CHEMECA-81, 9th Australasian Conference on Chem. Engng. paper A2.3, Christchurch, N.Z. (1981), pp. 65-72.

13. **Claflin, J.K. and A.G. Fane**, "Spouting with a Porous Draft Tube", Can. J. Chem. Eng. (1983) 61, 356-363.

14. **Khoe, G.K. and J. Van Brakel**, "Drying Characteristics of a Draft Tube Spouted Bed", Can. J. Chem. Eng. (1983) 61, 411-418.

15. **Grbavcic, Z.B., R.V. Garic, D.V. Vukovic, Dz.E. Hadzismajlovic and H. Littman**, "Comparative Characteristics of Spouted and Spouted Beds with a Draft Tube, 9th Int. CHISA '87 Congress, Paper E9.26, Prague, Czechoslovakia (1987).

16. **Brereton, C.M.H., N. Epstein and J.R. Grace**, "Gas Flow Distribution in the Annulus of Spouted Beds", Proc. 33rd Canadian Chem. Eng. Conf., Toronto, Vol. II (1983), pp. 785-790.

17. **Berruti, F., J.R. Muir and L.A. Behie**, "Solids Circulation in a Spout-Fluid Bed with Draft Tube", Can. J. Chem. Eng. (1988) 66, 919-923.

18. **Matthew, M.C., M.H. Morgan, III and H. Littman**, "Study of the Hydrodynamics within a Draft Tube Spouted Bed System", Can. J. Chem. Eng. (1988) 66, 908-918.

19. **Morgan, M.H., III and H. Littman**, "General Relationships for the Minimum Spouting Pressure Drop Ratio, PmS/PmF and the Spout-Annulus Interfacial Condition in a Spouted Bed", in "Fluidization", J.R. Grace and J.M. Matsen (Eds.), Plenum Press, New York (1980), pp. 287-296.

20. **Day, J-Y, H. Littman, M.H. Morgan, Z.B. Grbavcic, Dz.E. Hadzismajlovic and D.V. Vukovic**, "An Axisymmetric Model for Fluid Flow in the Annulus of a Spout Fluid Bed", Chem. Eng. Sci. (1991) 46 773-779.

21. **Yang, W.C. and D.L. Keairns**, "Studies on the Solids Circulation Rate and Gas Bypassing in Spouted Fluid-Bed with a Draft Tube", Can. J. Chem. Eng. (1983) 61, 349-355.

22. **Hadzismajlovic, Dz.E., D.S. Povrenovic, Z.B. Grbavcic, D.V. Vukovic and H. Littman**, "A Spout-Fluid Bed Drier for Dilute Solutions Containing Solids", Fluidization VI, Ed. J.R. Grace, L.W. Shemilt and M.A. Bergougnou, Engineering Foundation, N.Y. (1989), 277-284.

23. **Grbavcic, Z.B., D.V. Vukovic, S.Dj. Jovanovic, R.V. Garic, Dz.E. Hadzismajlovic, H. Littman and M.H. Morgan III**, "Fluid Flow Pattern and Solids Circulation Rate in a Liquid Phase Spout-Fluid Bed With Draft Tube", Can. J. Chem. Eng. (1992) 70, 895-904.

24. **Grbavcic, Z.B., R.V. Garic, D.V. Vukovic, Dz.E. Hadzismajlovic, H. Littman, M.H. Morgan III and S.Dj. Jovanovic**, "Hydrodynamic Modeling of Vertical Liquid-Solids Flow", Powder Technol. (1992) 72, 183-191.

25. **Littman, H., M.H. Morgan III, J.D. Paccione, S.Dj. Jovanovic, and Z.B. Grbavcic**, "Modeling and Measurement of the Effective Drag Coefficient in Decelerating and Non-Accelerating Turbulent Gas-Solids Dilute Phase Flow of Large Particles in a Vertical Transport Pipe", Powder Technology (1993) 77, 267-283.

26. **Littman, H., M.H. Morgan III, S.Dj. Jovanovic, J.D. Paccione, Z.B. Grbavcic and D.V. Vukovic**, "Effect of Particle Diameter, Particle Density and Loading Ratio on the Effective Drag Coefficient in Steady Turbulent Gas-Solids Transport", Powder Technology (1995) 84, 49-56.

27. **Littman, H., M. H. Morgan III and J. D. Paccione**, "A Pseudo-Stokes Representation of the Effective Drag Coefficient for Large Particles Entrained in a Turbulent Airstream", Powder Technology (1996) 87, 169-173.

28. **Littman, H.**, "The State of Development of the Draft Tube Spout-Fluid Bed Contactor for Large Particles", J. Serb. Chem. Soc. (1996) 61, 211-231.

29. **Morgan, C.B. and M.H. Morgan III**, "Statistical Analysis of a Spouted Bed Coater: Development of Simple Design Equations", paper presented at the 1986 American Statistical Association meeting, Chicago, IL.

30. **Morgan, C.B. and M.H. Morgan III**, "Using the Gamma Distribution to Model the Behavior of Engineering Systems", paper presented at the 1988 American Statistical Association meeting, Anaheim, CA.

31. **Mann, U., E.J. Crosby and M. Rabinovich**, "Number of Cycles Distribution in Circulating Systems", Chem. Eng. Sci. (1974) 53, 761-765.

32. **Mann, U. and E.J. Crosby**, "Cycle Time Distribution Measurements in Spouted Beds", Can. J. Chem. Eng. (1975) 53, 579-581.

33. **Mann, U.**, "Analysis of Spouted Bed Coating and Granulation. 1. Batch Operation", Ind. Eng. Chem. Process Des. Dev. (1983) 22, 288-293.

34. **Hahn, G.J. and S.S. Shapiro**, "Statistical Models in Engineering", Wiley, N.Y. (1967).

Table 1

Dimensions of the DTSFB of the Coater

Section	Dimensions	
Annulus	D_c = 152.4 mm	H_a = 457.2 mm
Draft tube	D_d = 76.2 mm	l_d = 914.4 mm
Spout inlet tube	d_i = 63.5 mm	L_i = 280 mm
Fountain tube	d_f = 150 mm	L_f = 1210 mm
Round to Square Expander	152.4 mm ID	bottom
Annulus to Vent Header	355.6 x 355.6 mm 304.8 mm	top height
Vent header	355.6 x 355.6 mm 609.6 mm	cross section height

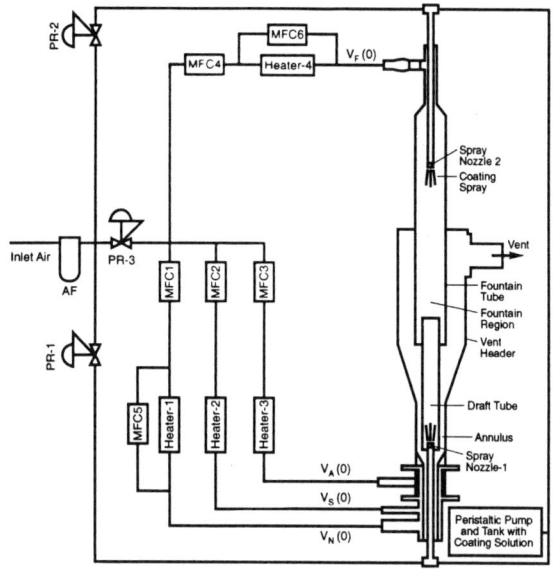

Figure 1 - Schematic diagram of coating system (1). Symbols defined in nomenclature section. PPs and TCs shown in (1).

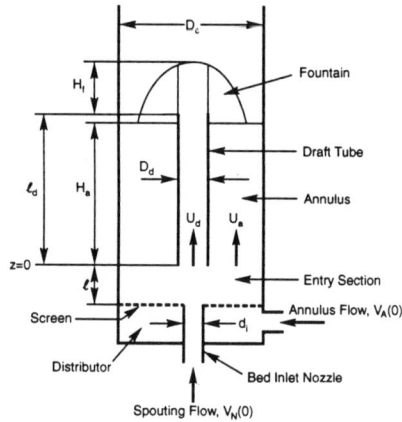

Figure 2 - Schematic diagram of a spout-filled fluid bed with a draft tube.

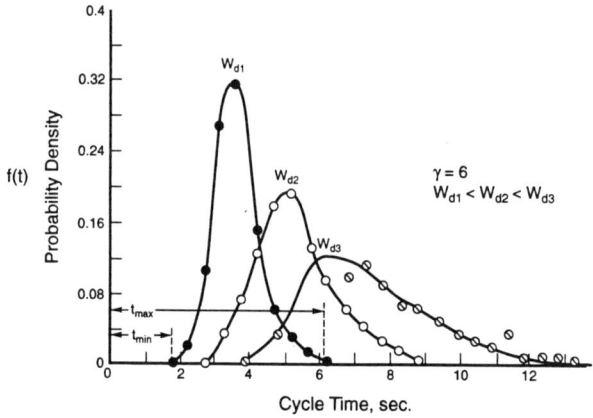

Figure 3 - CTD for three different flow rates using experimental data of Mann et al.

Characteristics of Particle Circulation in a Spouted Bed with a Draft Tube

Hongshen Ji, Atsushi Tsutsumi and Kunio Yoshida
Department of Chemical System Engineering, The University of Tokyo
7-3-1 Hongo, Bunkyo-ku, Tokyo 113, Japan

A spouted bed of 0.205 m diameter with a draft tube has been used to study the effect of gas velocity on the behavior of solids circulation. The solids circulation rate and gas bypassing were measured, and an optical fiber probe was developed to measure the fluctuation in solids flow associated with the measurement of pressure fluctuation. The power spectral density analysis of the time series data of the optical fiber probe and pressure fluctuation was performed to understand the solids circulation mechanism in conjunction with visual observation using a borescope.

Spouted bed technology has been applied to various industrial processes such as drying, coating and granulating solids [1]. A spouted bed with a draft tube (DTSB) is a modification to the conventional spouted bed and has the advantage of having more design flexibility and easy controllability. In addition, the minimum spouting velocity of the DTSB is less than that of the conventional spouted bed and its maximum spoutable bed height is limited only by the energy of the gas stream entering the bottom of the draft tube.

For the DTSB, the solids circulation rate is an important variable in predicting the performance of the bed, which is mainly dependent on the gas bypassing characteristics. Several investigators have studied the effects of operating and design parameters on solids circulation and gas bypassing [2,3,4,5,6]. The solids circulation rate G_s increases with increasing gas inlet velocity to reach a maximum and subsequently decreases gradually [2,3,7,8]. In the higher gas inlet velocity region, G_s appears to be steady [6,8,9].

Despite the accumulation of data on solids circulation rates in DTSBs, information on the mechanism of solids circulation remains limited. Yang and Keairns [10] observed the solids circulation in a semicircular DTSB by using high speed movies and found that solid transport in the draft tube was a slugging type. In addition, they observed that the frequency of slugging was between 4 to 8 Hz and inversely proportional to the total gas flow rate. Muir et al. [6] observed the entrainment of particles at the spout zone in a semicircular DTSB using a video camera. Their study showed that the majority of particles are entrained in clusters very close to the spouting gas inlet. The solid particles cross into the gas jet, cover the inlet and are carried away by the spouting gas, thus forming a particle cluster. The frequency of the cluster (8 to 12 Hz) was observed to increase with gas velocity. They stated that the solids circulation rate stems from jetting occurring in the entrainment region.

In this study, the solids circulation mechanism and the motion of clusters in a cylindrical DTSB were characterized by the spectral analysis of pressure fluctuation associated with direct visual observation using a borescope. In addition, an optical fiber probe was used to detect the individual particle frequency in the draft tube. Power spectral density analysis of these time series data was performed to understand the behavior of solids circulation.

EXPERIMENTAL

A transparent acrylic resin column of 0.205 m ID and 7 m height was used with a cylindrical draft tube of 5 cm ID and 0.8 m height, as shown in Figure 1. The inlet nozzle diameter and cone angle were 3.15 cm and 60 degrees respectively. The distance between the inlet nozzle and the base of the draft tube was set at 7.5 cm. The solid particles used in this experiment were glass beads with a mean particle diameter of 1 mm and a density of 2500 kg/m^3. Air was used as the spouting fluid and its flow rate was measured using an orifice

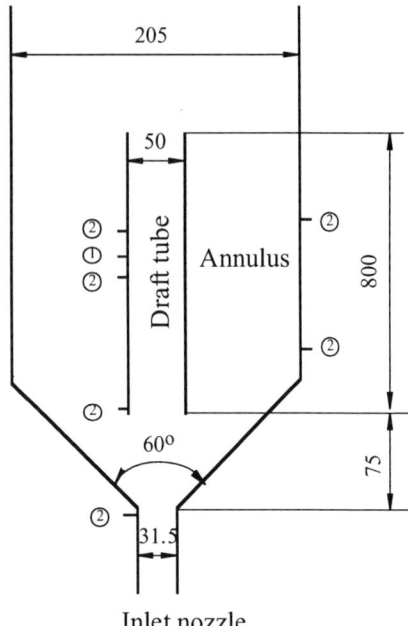

Figure 1. A schematic diagram of the experimental apparatus: (1) optical fiber probe, (2) pressure taps.

flow meter. The superficial gas velocity (U_g) based on the cross-sectional area of inlet nozzle was varied in the range of 20 to 74 m/s.

The particle velocity in the annulus U_{as} was determined by measuring the time required for a colored tracer particle to travel downward a constant distance. Assuming that the voidage in the annulus ε_a is the same as that in the packed bed, the solids circulation rate can be obtained by

$$G_s = (1 - \varepsilon_a)\rho_p U_{as} A_a \quad (1)$$

where A_a is the cross sectional area of the annulus and ρ_p is particle density. The gas velocity in the annulus was calculated from the Ergun equation [11] by measuring the static pressure drop [6].

The pressure fluctuation was measured by a pressure transducer (Setra Systems, C-239) in the draft tube and the entrainment region.

An optical fiber probe was developed to measure the particle frequency in the draft tube. The optical fiber probe consists of three fibers (0.2 mm diameter): one to transmit laser light and two to collect reflected light. The light reflected by the glass beads was converted into a voltage signal by a phototransistor. The optical fiber probe was mounted horizontally and located in the center axis of the draft tube at the middle section.

The voltage signals of pressure fluctuation and particle frequencies were recorded on the data recorder and digitized with an AD converter at a sampling rate of 1 ms. A typical sample size was 32760 points. A borescope (Olympus, K27-18-00-62) was used to observe the motion of clusters in the entrainment region and the draft tube.

RESULTS AND DISCUSSION

Figure 2 shows the effect of U_g on G_s. In this system, the minimum spouting gas velocity (U_{ms}) was 19.9 m/s. With the increase in U_g, the solids circulation rate increases to reach a maximum at the gas velocity of 37 m/s (hereafter called U_g^*) and subsequently decreases. When U_g is larger than 60 m/s (hereafter called U_{mp}), G_s becomes constant. From the visual investigation using the borescope in the entrainment region, it was observed that solid particles move from the annulus dense region into the gas jet intermittently and are then entrained upward in the draft tube by the spouting gas, forming clusters. A similar phenomenon has been reported in the visual observation using a semicircular DTSB by Muir et al. [6]. It was found that in the high gas velocity region ($U_g > U_{mp}$) no distinct clusters were observed in either the entrainment region nor the draft tube.

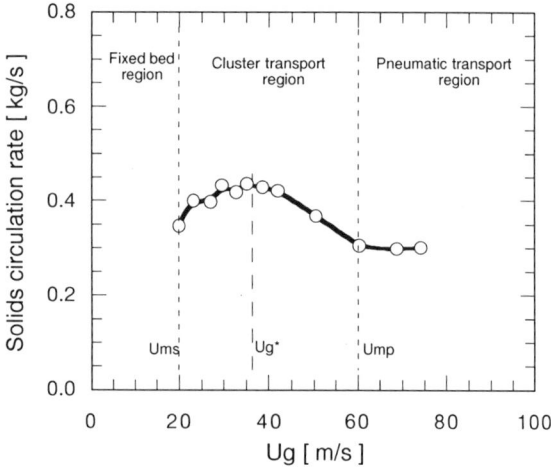

Figure 2. Effect of gas velocity on solids circulation rate

The gas bypassing results obtained from measurement of the pressure drop in the annulus are shown in Figure 3. It can be seen that, as U_g increased, the superficial gas velocity in the draft tube (U_d) increased linearly. The fraction of gas flowing through the annulus (Q_a/Q_g) decreased with U_g.

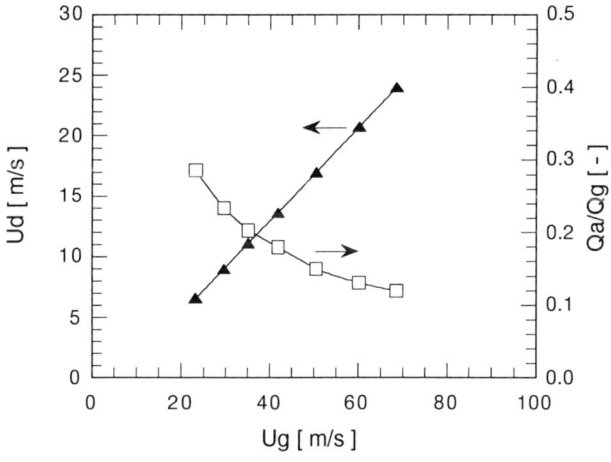

Figure 3. Effect of gas velocity on gas bypassing

A typical output signal for pressure fluctuation in the entrainment region of the DTSB is shown in Figure 4. It can be seen that the pressure fluctuation exhibits a periodic nature. Figure 5 shows the power spectral density function of the pressure fluctuation at a gas velocity of 35.1 m/s. The power spectral density exhibits two sharp peaks near 7 and 50 Hz. The frequency component near 50 Hz is due to the blower because the empty column has a dominant frequency of 50 Hz in the pressure fluctuation independently of gas velocity. On the other hand, it can be considered that the peak near 7 Hz is attributable to the frequency of cluster formation in the entrainment region. This is

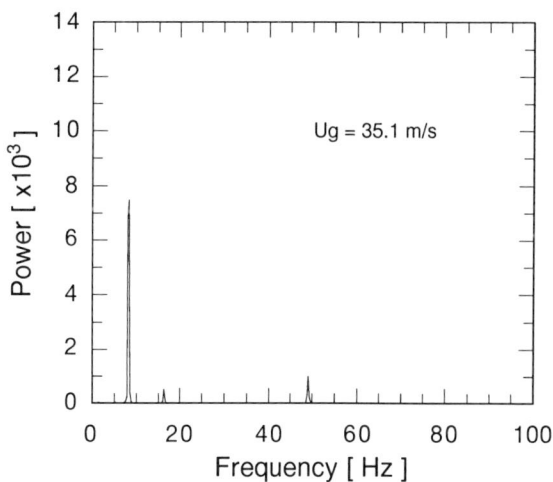

Figure 5. Power spectrum of pressure fluctuation in the entrainment region

confirmed by frame-by-frame analysis of the video tape of the borescope image, where it was observed that the formation of clusters takes place at frequencies ranging from 7 to 10 Hz.

The effect of gas velocity on the frequency of cluster formation and its power spectral density are shown in Figures 6 and 7, respectively. As can be seen in Figure 6, the frequency of cluster formation increases slightly with the gas velocity. When the gas velocity is

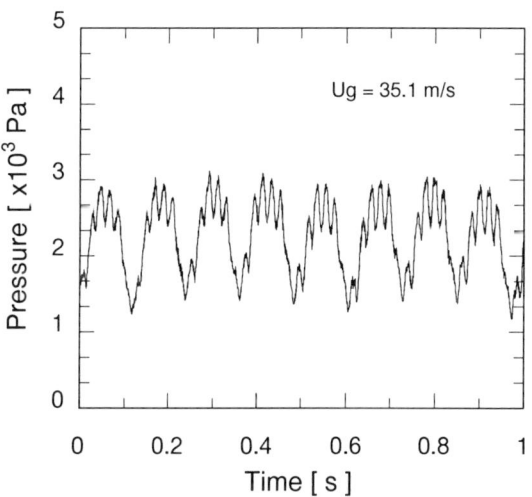

Figure 4. Representative signals of pressure fluctuation in the entrainment region

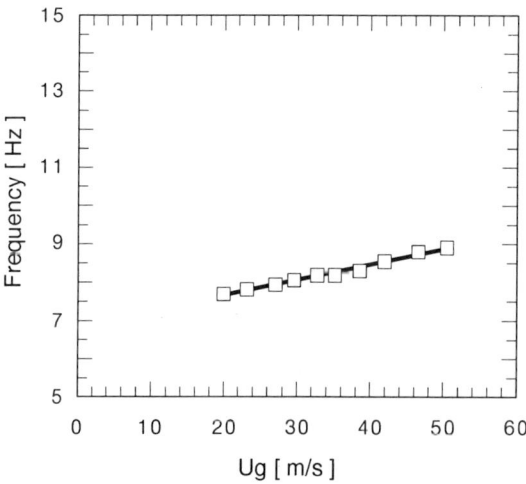

Figure 6. Effect of U_g on cluster frequencies in the entrainment region obtained by the pressure fluctuation

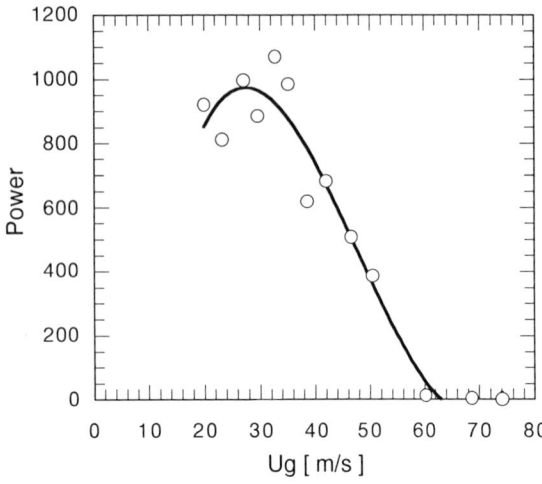

Figure 7. Effect of Ug on the power spectral density of the dominant frequency in the entrainment region

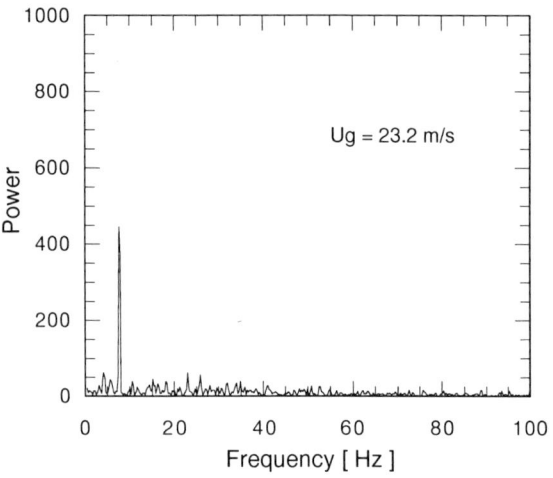

Figure 8. Power spectrum of signal of optical fiber probe in the draft tube

higher than U_{mp}, there is no dominant peak in the spectrum of pressure fluctuation. With increasing gas velocity, the power spectral density increases between U_{ms} and U_g^*, and decreases linearly above U_g^* as can be seen in Figure 7.

As the gas velocity increases, the gas velocity in the draft tube increases linearly. This enhances the entrainment of solid particles, resulting in increasing cluster size as well as cluster frequency. This leads to an increase in the solids circulation rate with gas velocity as shown in Figure 2. After the solids circulation rate reaches its maximum at U_g^*, the power spectral density drops rapidly. This implies that the solids flow in the draft tube becomes more homogeneous and dilute. As the gas velocity reaches to U_{mp}, the solids flow fluctuations damp, reducing the formation of clusters. The disappearance of the peak near 7 Hz in the power spectrum above U_{mp} indicates that particles are dispersed homogeneously and are carried upward in the draft tube as a continuous dilute flow.

Figure 8 shows the power spectral density function of particle frequency measured by the optical fiber probe in the midsection of the draft tube. The spectrum has one distinct peak at about 7 to 10 Hz, indicating the periodical motion of particles. The frequency of this peak is in fair agreement with that of the pressure fluctuation in the draft tube.

Figure 9 shows the effect of gas velocity on the

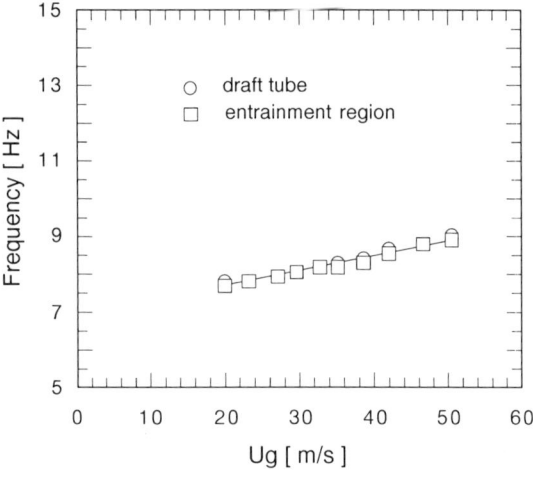

Figure 9. Effect of Ug on the cluster frequencies

dominant frequencies in the draft tube and the entrainment region. It was found that for all gas velocities the frequency in the draft tube is same as that in the entrainment region. This reveals that the clusters formed in the entrainment region enter the draft tube and are carried upward without breakdown. This accords with the result of visual observation in the draft tube using the borescope. It can be concluded, therefore, that the dominant mechanism of the particle circulation in a DTSB is a heterogeneous slugging-type transport with the periodical formation of clusters with the frequency of 7 to 10 Hz. At high gas velocities above U_{mp}, a continuous stream of dilute solid suspension passes through the draft tube without the formation of clusters, leading to the stable pneumatic conveying of solids.

CONCLUSIONS

The solids circulation in a cylindrical spouted bed of 0.205 m diameter with a draft tube of 5 cm diameter has been studied based on visual observation using a borescope and the analyses of pressure fluctuation and the optical signals. The power spectral density functions in both the entrainment region and the draft tube exhibit a sharp peak with a frequency of 7 to 10 Hz which is attributable to the motion of clusters. In the entrainment region, solid particles move from the annulus into the gas jet to form clusters periodically and are carried upward in the draft tube. With increasing gas velocity the solids circulation rate increases to reach a maximum at U_g^*. This is associated with an increase in the size and frequency of clusters. Above U_g^* the power spectral density drops rapidly, indicating that the solids flow becomes homogeneous and dilute. When the gas velocity is larger than U_{mp}, particles are dispersed homogeneously and transported upward without the formation of clusters, leading to the stable pneumatic conveying of solids.

NOTATION

A_a cross sectional area of the annulus, m^2
G_s solids circulation rate, kg/s
Q_a gas flow rate in the annulus, m^3/s
Q_g total inlet gas flow rate, m^3/s
U_{as} particle velocity in the annulus, m/s
U_d superficial gas velocity in the draft tube, m/s
U_g superficial nozzle inlet gas velocity based on the nozzle inlet cross sectional area, m/s
U_g^* superficial nozzle inlet gas velocity where G_s reaches a maximum, m/s
U_{ms} minimum spouting gas velocity, m/s
U_{mp} minimum pneumatic transport gas velocity, m/s

Greek letters

ε_a voidage in the annulus, -
ρ_p particle density, kg/m^3

LITERATURE CITED

1. Mathur, K.B. and N. Epstein, "Spouted Beds", Academic Press, New York, pp. 187-236 (1974).

2. Claflin, J.K. and A.G. Fane, "Spouting with a Porous Draft-tube", *Can. J. Chem. Eng.*, **61**, 356 (1983).

3. Ferreira, M.C. and J.T. Freire, "Fluid Dynamics Characterization of a Pneumatic Bed Using a Spouted Bed Type Solid Feeding System", *Can. J. Chem. Eng.*, **70**, 905 (1992).

4. Kalwar, M.I., G.S.V. Raghavan and A.S. Mujumdar, Circulation of Particles in Two-dimensional Spouted Beds with Draft Plate", *Powder Tech.*, **73**, 233 (1993).

5. Khoe, G.K., and J.van Brakel, "A Draft Tube Spouted Bed as Small Scale Grain Dryer", *I. Chem. E. Symp. Ser.*, No**59**, 6:6/1(1980).

6. Muir, J.R., F. Berruti, L.A. Behie, "Solids Circulation in Spouted and Spout-fluid Beds with Draft-tube", *Chem. Eng. Comm.*, **88**, 153 (1990).

7. Buchanan, R.H., and B. Wilson, "The Fluid-lift Solids Recirculator", *Mech. Chem. Eng. Trans.*, May 117(1965).

8. Ichiji, K., "Fundamental Characteristics of Fluidized Beds and Spouted Beds with Draft Tubes" Ph.D. Thesis, University of Tokyo, Japan (1991).

9. Berruti, F., J.R. Muir and L.A. Behie, "Solid Circulation in a Spout-fluid with Draft tube", *Can. J. Chem. Eng.*, **66**, 919 (1988).

10. Yang, W.C., and D.L. Keairns, "Design of Recirculating Fluidized Beds for Commercial Applications", *AIChE Symp. Ser.*, No.176, **74**, 218 (1978).

11. Ergun, S., "Fluid Flow Through Packed Columns", *Chem. Eng. Prog.*, **48**, 89 (1952).

Use of the Attenuation of Acoustic Pulsed Waves for Concentration Measurement in Gas-Solid Pipe Flow

Stephen Tallon and Clive E. Davies
Industrial Research Ltd., P.O. Box 31-310, Lower Hutt, New Zealand

The flow rate of solids in a pneumatic conveying line can be measured on-line using acoustic waves in the audible frequency range. A prototype measurement instrument is described using four microphones attached to pressure taps along the wall of the pipeline. Conveying velocities were calculated from the difference between upstream and downstream propagation velocities and the solids volumetric concentration was correlated with the attenuation of the signal. A simple on-line calibration method using injection of a known mass of material was developed, and it fitted well to steady state test data over a wide range of conveying velocities. The system was tested in a three inch horizontal pipeline conveying 150 μm mean diameter sand particles in the dilute phase regime.

The solids flow rate in pneumatic conveying pipelines is a variable which has proved to be very difficult to measure on-line with a system which is accurate, reliable, and non-intrusive. Many of the difficulties encountered arise from the complex nature and distribution of the phases in two phase gas-solid pipe flow, and from the harsh conveying environment. A number of alternatives to this problem have been investigated but to the authors' knowledge no flow rate measurement instrument has been developed which has widespread commercial success.

Acoustic waves provide a viable method for making this measurement. Acoustic waves in the low to audible frequency range exist naturally in any conveying system, for example generated by the air mover or the solids motion, and these have been used in previous work as the basis for a solids flow rate measurement instrument [1]. The method measures the conveying velocity from its effect on the propagation velocity, and the solids concentration is correlated with attenuation of the wave along the pipe. This same approach can also be applied using locally introduced waves [2]. Locally introduced waves have the advantage of a controllable frequency, avoidance of some resonance problems encountered with natural continuous waves,

---

Industrial Research Limited, Gracefield, Lower Hutt, New Zealand

and the ability to make a velocity measurement which is independent of the acoustic propagation velocity.

Successful application of a flow rate measurement instrument depends on its ability to be accurately and easily calibrated. This paper examines the performance of a prototype instrument using an introduced pulsed acoustic wave, and tests a calibration method which is based on injection of a known quantity of solids.

METHOD

The measurement system is shown in Figure 1, and consists of a centrally located acoustic source between two pairs of microphones. The generated sound wave passes in both directions along the pipe and its passage is detected as it passes the transducers. The conveying velocity and the solids volumetric concentration are each calculated independently from these measurements. From the velocity and the solids concentration the solids mass flow rate can then be calculated.

Velocity Measurement

By introducing a pulse of sound through the speaker, the velocity of the wave in both directions along the pipe can be measured by cross correlation of the signals from the microphones. The sound velocity in a moving medium is equal to the acoustic velocity of the medium plus the velocity at which the medium is travelling [3], see Equation (1), where v^+ is the downstream velocity of the wave, v^- is the upstream velocity, v_S is the acoustic velocity of the medium, and v_C is the convective velocity. Consequently the convective velocity of the medium is given by half the difference between the upstream and downstream velocities, see Equation (2). The solids velocity in a suspended pipe flow is generally a little lower than the air velocity, but for dilute flows the sound propagates through the continuous gas phase so that the measured convective velocity from Equation (2) is that of the gas phase. For very dilute flows this can also be assumed to be equal to the superficial air velocity. The acoustic velocity of the medium can also be calculated and is given by the mean of the upstream and downstream velocities, see Equation (3).

$$v^+ = v_S + v_C, \quad v^- = v_S - v_C \quad (1)$$

$$v_C = \frac{v^+ - v^-}{2} \quad (2)$$

$$v_S = \frac{v^+ + v^-}{2} \quad (3)$$

Concentration

The concentration can be correlated against its attenuating effect on the generated signal. The strength of the signal can be measured by calculating the peak covariance [4] between the input signal sent to the speaker, and the signal recorded at each of the four microphones, see Equation (4). Here $\gamma_{I,j}(k)$ is the covariance between the input, I, and point j at a lag of k points. $P_{a,t}$ is the recorded pressure signal at point a, and time t, $\overline{P}$ is the mean value of P, and n is the total number of points in the series.

$$\gamma_{I,j}(k) = \frac{1}{n}\sum_{t=1}^{n-k}(p_{I,t} - \overline{p}_I)(p_{j,t+k} - \overline{p}_j) \quad (4)$$

Assuming a Lambert-Beer type law applies [5] to the decay of the signal with distance, Equation (5), then the solids volumetric concentration can be written as a function of the measured variables, $\gamma_{I,1}$, $\gamma_{I,0}$. Here $\alpha(c)$ is the attenuation coefficient for solids concentration c, and l is the distance the wave travels through the medium. The attenuation can be measured in either direction along the pipe, but the value of the length, l, is a function of both the direction and the convective velocity. If the sound is travelling in the same direction as the conveyed medium, then the distance the wave travels through the suspension between the two microphones is less than the actual physical distance between the transducers, and vice versa for the upstream wave, as indicated in Equation (6). Here x is the physical distance between the two transducers.

$$\gamma_{I,1} = \gamma_{I,0} \cdot e^{-\alpha(c).l} \quad (5)$$

$$l_\pm = x \cdot \frac{v_S}{(v_S \pm v_C)} \quad (6)$$

Calibration

The system can be calibrated by injecting a known quantity of the conveyed material into the pipeline and assuming a relationship between the attenuation coefficient and the solids concentration. This relationship has in previous work been found to be a linear one, ie $\alpha(c) = K.c$. Using this with Equations (5),(6) and Equation (7) gives the solids mass flow rate in terms of the measurable signal strengths, Equation (8). $\dot{M}$ is solids mass flow rate, v_{Solid} is the solids velocity, A is the pipe area, and ρ_{Solid} is the solids particle density

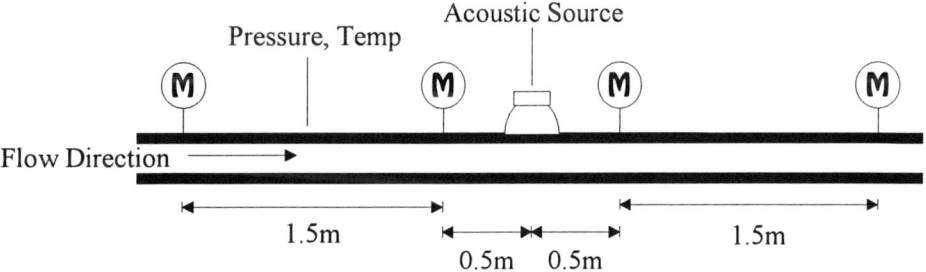

Figure 1 - Schematic drawing of measurement system

$$\dot{M} = c.v_{Solid}.A.\rho_{Solid} \qquad (7)$$

$$\dot{M} = \frac{-1}{K}.\frac{\ln(\gamma_{1,1}/\gamma_{1,0})}{x.v_s/(v_s \pm v_c)}.v_{Solid}.A.\rho_{Solid} \qquad (8)$$

Integrating this expression over the time that the known quantity of material is injected into the system gives an expression for the total mass of material in terms of the parameter K, thus giving a calibration. Calibration points generated at a number of different operating conditions can be used to confirm the appropriateness of the attenuation model chosen. This calibration can then be compared with calculated values of the solids mass flow rate under known steady state operating conditions.

EXPERIMENTAL

The system used is shown schematically in Figure 2. It is made from 76.2 mm outside diameter 304 stainless steel tube and the air is supplied by a Rootes type blower. The air flow rate was measured by a proprietary velocity head measurement instrument, a Diamond II Annubar from Deitrich Corp., USA. A continuous flow of solids could be fed through a calibrated orifice and rotary valve feeder, or for calibration purposes, a known mass of material could be injected into the pipeline with the blowtank.

Four microphones were used, which had a high sensitivity and a resonant frequency of 13.6 kHz, and were arranged as in Figure 1. These were logged at a rate of 25000 Hz per channel. The introduced sound was a composition of 1000, 1330, and 1900 Hz sine

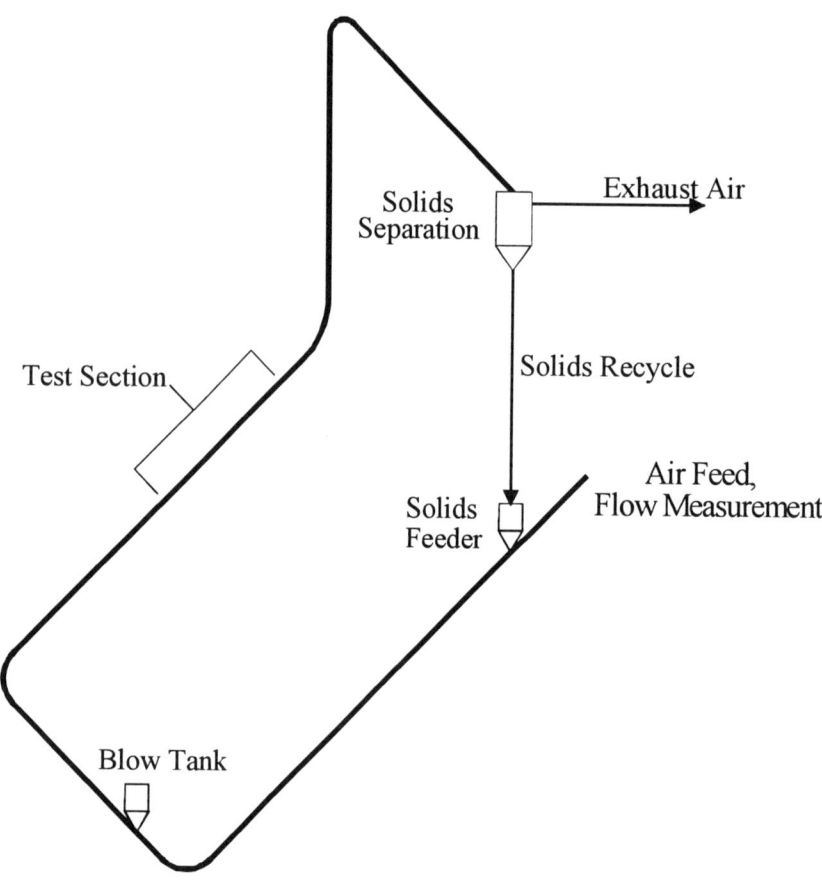

Figure 2 - Schematic drawing of conveying circuit

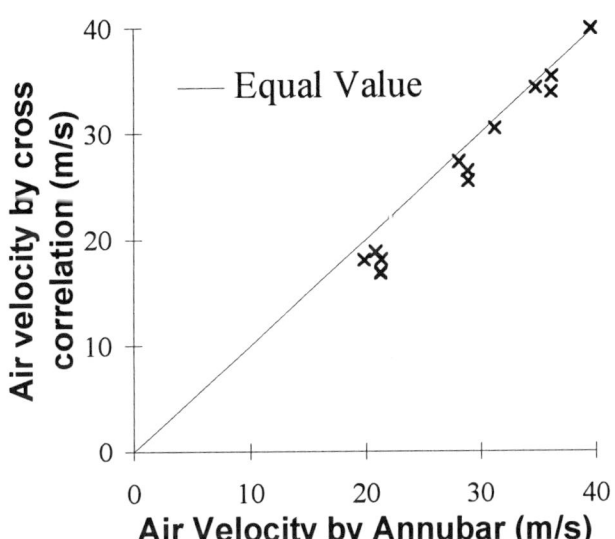

Figure 3 - Comparison of calculated air velocities with the Annubar measurement.

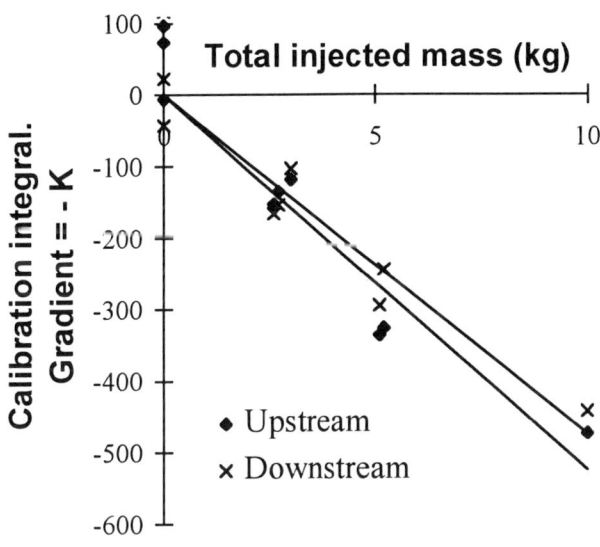

Figure 4 - Calibration curve. Y-axis is:
$$\sum \frac{\ln(\gamma_{I,1}/\gamma_{I,0})}{x.v_s/(v_s \pm v_c)}.v_{Solid}.A.\rho_{Solid}.dt$$

waves generated for 1/100th of a second at a rate of 10 per second. Measurements were taken for 25 seconds under a range of steady state operating conditions including air velocities from 20 to 40 m.s^{-1} and a solids mass loading ratio up to 5:1. Calibration pulses of up to 10kg of material were injected over the same range of air flow rates and at solids flow rates similar to those from the rotary valve feeder. Gauge pressure and temperature measurements were also taken for each flow condition.

RESULTS AND DISCUSSION

Figure 3 shows the air velocity calculated by cross correlation of the acoustic waves, compared to that measured by the Annubar with an allowance made for the backflow of air through the rotary solids feeder. The Annubar measurement returns higher values by up to about 7% at the lower velocities but has a number of uncertainties with it. Work is currently under way to install an accurate standard flow supply for comparison with the velocities measured by the acoustic technique.

Figure 4 shows the result of integrating Equation (8) and plotting against the total mass injected. The calibration coefficient, K, is calculated from a linear regression through the points. Two curves are shown, representing the attenuation of the input wave in the upstream and downstream directions. The downstream section has a lower attenuation relationship than the upstream section by about 10%. There is no immediate explanation for this difference but it has been attributed to the proximity of the test section to the next bend in the pipeline. The section that the downstream attenuation was measured over was less than 1m from the bend and is sufficiently close for a reflected wave from the bend to interfere with the original signal. A measurement section more than 2m from the bend is not expected to show this effect.

The calibration was then applied to the recordings taken under steady state operation with the solids fed by the rotary valve feeder. The comparison of calculated solids mass flow rates with the known flow is given in Figure 5. There is some deviation at the highest solids flow rate with changes in the air flow rate. The point on the line is at 20 m.s^{-1} and the other points below the line are for velocities 32 m.s^{-1} and 39 m.s^{-1}. The lowest point is about 15% lower than the actual solids flow rate. The measurements at lower solids concentrations however are more accurate and repeatable. The effect of the distribution of solids over the cross section of the pipeline is expected to be more significant at higher solids loading and lower conveying velocities where the solids tend to settle more on the bottom of the pipe and form dense clusters or strands. At low concentrations the relationship between signal strength and concentration is reasonably linear, but for

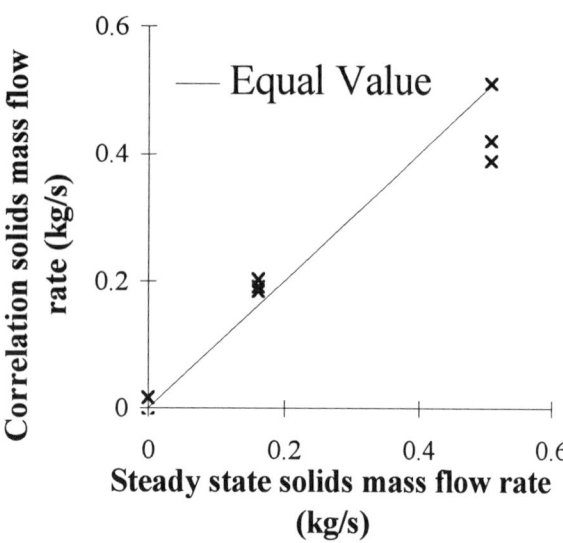

Figure 5 - Calculated solids mass flow rate compared to steady state flowrate through calibrated orifices. Air velocity from 20 to 40 ms^{-1}.

very dense local concentrations nonlinearities can become significant.

The solids velocity in this work is estimated from the measured air velocity by assuming a slip velocity, which is subject to some variation with solids loading. Direct measurement of the solids velocity, using the same transducer layout, is the subject of current work and may lead to accuracy improvements. It is also possible to measure the attenuation over a longer pipe length to improve the resolution of low solids concentrations.

CONCLUSIONS

A calibration technique based on the injection of a known batch of solids from a blow tank has been tested for the calibration of a solids flow meter for use in pneumatic conveyors. The calibration proved to be accurate except for some deviation with changing air flow rate at the highest solids concentrations tested. The assumption of a linear relationship between the attenuation coefficient and solids volume concentration has been justified.

NOTATION

a	Index representing signal source (-)
A	Pipe cross sectional area (m^2)
c	Solids volumetric concentration (-)
dt	Sampling period (s)
I	Index for input signal sent to speaker (-)
j	Index for microphones (-)
k	Time lag in units of the sampling period (-)
K	Calibration constant, Equation (8), (m^{-1})
l	Propagation distance (m)
$\dot{M}$	Solids mass flow rate (kg.s^{-1})
n	Number of points in sample (-)
$P_{a,t}$	Pressure at point a, and time t(Pa^2)
$\overline{P}$	Mean value of P (Pa^2)
t	Time index in units of the sampling period (-)
x	Distance between transducers (m)

Greek Symbols

$\alpha(c)$	Attenuation coefficient (m^{-1})
ρ_{Solid}	Solids particle density (kg.m^{-3})
v^-	Sound velocity in upstream direction (m.s^{-1})
v^+	Sound velocity in downstream direction (m.s^{-1})
v_C	Gas phase convective velocity (m.s^{-1})
v_S	Velocity of sound (m.s^{-1})
$\gamma_{I,0}$	Maximum covariance between input and pressure signal at reference point 0 (Pa^2)
$\gamma_{I,j}(k)$	Covariance between input I and point j at a lag of k (Pa^2)
$\gamma_{I,l}$	Maximum covariance between input and pressure signal at a distance l from reference point 0 (Pa^2)
v_{Solid}	Solids Velocity (m.s^{-1})

LITERATURE CITED

1. **Tallon, S., Davies, C.E.** "Flow Rate Monitoring and Measurement in Dilute Phase Pneumatic Conveying using Pressure Fluctuations," AIChE symposium series volume on Fluidization and Fluid Particle Systems, 91, 308, 137, (1995).
2. **Tallon, S., Davies, C.E.** "Flow velocity measurement in pneumatic conveyors using pulsed sound waves", Presented at the 5th World Congress of Chemical Engineering, San Diego, 14-18 July, (1996).
3. **Munjal, M.L.** "Acoustics of ducts and mufflers," John Wiley and Sons Inc., New York (1987).
4. **Bendat, J.S., Piersol, A.G.** "Random Data - Analysis and Measurement Procedures," John Wiley and Sons Inc., New York (1986).
5. **Iinoya, K., Masuda, H., Watanabe, K.** "Powder and Bulk Solids Handling Processes", Marcel Dekker Inc., New York (1988).

Characterization of Flow Properties of Very Fine Powders at Ambient and Elevated Temperature
[A Novel Experimental and Theoretical Approach]

H.O. Kono, L. Richman, and J. Su
Department of Chemical Engineering, West Virginia University, Morgantown, WV 26506

D. Smith
US-DOE, Morgantown Energy Technological Center, Morgantown, WV 26505

A new experimental method was developed to measure the rheological parameters, tensile strength (δ_f) and plastic deformation coefficient (Y), of fine or ultra fine powder at ambient and elevated temperature. This method was proven to be viable by comparing the results to the established split-cell method for determining the tensile strengths of powders. The split-cell method can only be used at ambient temperature and only applicable for powders within a certain range of interparticle force. Combining the measurement methods of the split-cell, fluidized bed, and the new packed bed method, the resulting rheological parameters were found to be significant.

Due to the recent rapidly growing application of various fine or ultra fine powders, the practical prediction and reliable measurement methods of flow properties of these fine powders have been very much requested. As reported by [1], [2], and [3], the flow properties of fine powders were found to be characterized by using characteristic rheological properties such as plastic deformation coefficient (Y) and tensile strength (σ_f) of fine powders under a well defined bulk powder condition. Using the fracture model of bulk fine powders by [1], we can predict the flow patterns of almost all fine powders, knowing the rheological parameters of Y and σ_f as shown in Figure 1. Therefore, the development of rheological parameter measurement for various fine powders will become very important.

Furthermore, it is requested by practical application to predict or measure the flow properties of fine powders not only at ambient temperature, but also at elevated temperature. With respect to size ranges of fine powders, it is requested to cover a very wide range of diameters, e.g., 100μ, 50μ, 10μ, 1μ, 0.1μ, and 0.01μ.

THEORETICAL AND EXPERIMENTAL

The fine powders of calcium carbonate and anhydrous calcium sulfate, which are both chemically very pure, with the primary particle size of 2-5μ, were used as sample powders. Filter cakes consisting of compacted bulk powder structures, where the concept of effective particle size (d*) will be introduced in the following discussion, should be characterized experimentally and theoretically.

The particle packing structure, according to our Unit Particle Layer Model, can schematically be shown in Figure 2, where the tensile strength (σ_f) can be defined by Rumpf's Equation [5]. The tensile strength, σ_f, can be expressed as Equation (1).

$$\sigma_f = \frac{1-\epsilon}{\pi} k \frac{H}{d^{*2}} \quad [Pa] \qquad (1)$$

Because this equation is qualitative, this equation was not used to get the tensile strength of σ_f. Therefore, we did not use Equation (1) to predict the σ_f values, but measured it by experiment. Nevertheless, it still provides basic characteristics of an agglomeration phenomenon. For example, during the flyash deposition on the ceramic filter, the compaction of the powder structure will start due to the hot gas flow across the filter cakes. With respect to the change of filter cake structure affecting the tensile strength, we can qualitatively consider that the higher the surface gas velocity, the greater the compression forces working on unit layers, the smaller the porosity (ϵ), the larger the coordination number (k), and thus the larger the fracture tensile strength (σ_f). With respect to the plastic deformation coefficient (Y) of bulk powders, we can say that the more flexible or softer the structures, the smaller the deformation coefficient (Y). Therefore, the value of Y is depending upon the range of strains. In this approach, we defined Y only within small ranges of strain to obtain only relative values of each powder.

Because of the existence of relatively strong forces in the case of fine powders (particle sizes smaller than 10μ), the effective particle size should be introduced, which could experimentally be determined as shown in later discussions. In this work, we had to prepare reproducible filter cake samples in the laboratory using the porosity of the cakes as a criterion of the filter cake quality. We prepared the filter cake samples in the porosity range of 40-71%, and measured the mechanical (i.e., rheological) properties of samples under the in-situ condition.

Now, it is requested to measure the properties of filter cakes not only at ambient temperature, but also at elevated temperature (300~500°C). With respect to the tensile strength measurement, the well-known conventional split cell method can only be used under ambient temperature conditions. We have developed a new approach to measure the rheological parameters (Y and σ_f) using the compaction and expansion of bulk fine powders under well-defined packed bed conditions. This new approach can be applied both at ambient and elevated temperature and also at ambient and high pressure. The combination of three rheological measurement methods can provide a comprehensive characterization approach to measure the rheological parameters. It consists of the following methods, i.e.,

(1) Well-known tensile strength measurement of compacted fine powders by split-cells at ambient temperature for a limited range of powder sizes, e.g., [7], and [3].
(2) Tensile strength and deformation coefficient measurement method of aeratable fine powders in smoothly aerated beds at ambient and elevated temperature, e.g., [2].
(3) Tensile strength and deformation coefficient measurement method of bulk, fine, very fine, or ultra-fine powders under a wide range of porosities at ambient and elevated temperature, which is proposed in this paper.

For the case of this project, we used Method (3) for high porosity filter cakes at ambient and elevated temperature, and Method (1) for medium porosity filter cakes at ambient temperature. The principle of the split-cell method, Method 1, is to be referred to the published papers already mentioned. The principle of the rheological parameter Method (3) was developed, using the gas flow through the porous structures of filter cakes so that the compression and expansion within the fine powder packed bed can smoothly be controlled.

The theoretical approach written in this section was developed by careful observation of many preliminary experiments to set up the most reasonable assumptions used in the following discussions. Therefore, most of the assumptions made were really supported by experiments. Fine powders are first put into the cylinder on a porous plate with the loosely packed bed height of H_{lp}, consisting of many unit layers with the equivalent particle height of $d_{lp}*$ as shown in Figure 2. The number of the unit layers is to be $N_{lp} = H_{lp}/d_{lp}*$. Then, a certain amount of downward gas flow is introduced into the vessel, and the bed is compressed to reach the compressed bed height of H_c with the equivalent unit layer height of d_c*. The gas pressure drop across the bed is ΔP_c* and that across the unit layer is ΔP_c** as shown in Figure 2. Thus, the packed bed with a certain designed amount of porosity (ϵ) can be obtained by controlling the gas velocity. Using the fine powder packed-bed bulk powders with the well-defined porosity, we can prepare the samples for the fracture test. Next, the gas starts flowing upwardly and the gas flow rate gradually increases to the flow rate point at which the initial fractures of the packed bed powder structures occur. The fracture starts at the bed height of H_f, or the unit layer height of d_f*. It was experimentally verified that the fracture can occur randomly with respect to the axial direction at a few locations simultaneously. Although we could measure the tensile strength (σ_f) and the plastic deformation coefficient (Y) experimentally, the definition of these rheological parameters and other related phenomena can theoretically be expressed as follows.

During the compression of powders from the loosely packed bed to the certain compressed packed bed, the gas flow pressure drop across the unit powder layer will reach to ΔP_c** per unit layer.

$$\Delta P_c^{**} = \Delta P_c^{*}/N_c \quad (2)$$

where ΔP_c* is the gas pressure drop across the bed height after the compression. $N_c = (H_c/d*_c)$ is the number of the unit layers at the compression stage, and approximately the same as N_{lp}, which is the number of the unit layers at the loosely packed stage. Hereby, H_c is the bed height after the compression of the bed has been accomplished by passing a certain gas flow rate.

$$N_{lp} = N_c \qquad (2a)$$

By introducing the effective diameter d^*_c of powders at the compressed stage, ΔP_c^{**} can be expressed by combining Ergun's Equation with Equation (2)

$$\Delta P_c^{**} = \frac{150\, H_c \dfrac{(1-\epsilon_c)^2}{\epsilon_c^3} \dfrac{\mu U_c}{(\varphi_s d_c^*)^2}}{N_c} \qquad (3)$$

where ϵ is the bed porosity, μ is the gas viscosity, U_c is the gas velocity, and ϕ_s is the shape factor of an equivalent diameter. The fracture strength (σ_f) of compressed filter cakes and the plastic deformation coefficient (Y) of the powder within the small range of strains can be expressed as Equation 4

$$\sigma_f = \Delta P_f^{**} = \frac{\Delta P_f^*}{N_f} \qquad (4)$$

$$Y = \frac{\Delta P_c^{**}}{\dfrac{H_{lp} - H_c}{H_{lp}}} = \frac{\dfrac{\Delta P_c^*}{N_c}}{\dfrac{\Delta H}{H_{lp}}} \qquad (5)$$

whereby ΔP_f is the gas pressure drop across the bed when the fracture occurs and N_f is the number of the unit layers at that time. The number of unit layers should remain constant during the processes of compression and fracture as shown in Equation (6).

$$N_f \approx N_{lp} \approx N_c \qquad (6)$$

With respect to N_f, these numbers were actually obtained by experiment. In general for any fine powder, there is a linear relation between the log σ_f and ϵ [4]. With respect to the $CaCO_3$ powder sample, we got almost identical results, showing that the log σ_f vs. ϵ characteristic curve is almost linear. Using these results, the effective particle diameter of powders of d_f^* could experimentally be obtained, which was approximately 600μ. This is in good agreement with microscopic observation of effective particle size. For the case of anhydrous calcium sulfate, the effective diameter was also found to be in good agreement with microscopic observation. Accordingly $N_f (= H_f/d_f^*)$ could experimentally be obtained. Combining with Ergun's Equation, Equation (3) can be expressed as follows:

$$\sigma_f = \Delta P_f^{**} = \frac{150\, H_f \dfrac{(1-\epsilon_f)^2}{\epsilon_f^3} \dfrac{\mu U_f}{(\varphi_s d_f^*)^2}}{N_f} \qquad (7)$$

Although we described Equations such as Equations (3) and (7), note that all the data presented in this study was obtained directly by experiments. These theoretical equations, e.g., (3) and (7) are only useful to understand the physical meaning of ΔP_c^{**} and ΔP_f^{**}.

EXPERIMENTAL RESULTS AND DISCUSSION

As sample fine powders, calcium carbonate and calcium sulfate fine powders (2-5μ in diameter) were used. The calcium carbonate powder was not affected by the humidity of the air, but the anhydrous calcium sulfate powder was sensitive against the moisture existing in the air. The sample powders were maintained at 300°C and then kept in a desiccator. The compression of sample powders ($CaSO_4$ and $CaCO_3$) could be characterized in Figure 3, showing the compression stress per unit layers (ΔP^{**}comp.) versus the porosity (ϵ) of the compressed powder cylinder. Because we used the gas flow to compress the bulk powder, we obtained very homogeneous powder structures of the samples at ambient and also at elevated temperature. In Figures 4a and 4b, the tensile strength (σ_f) of compressed sample powders ($CaSO_4$ and $CaCO_3$) are shown as the functions of porosities of sample powders at ambient and elevated temperatures. These fracture tests were carried out by flowing gas from the bottom of the powder to the top until the fracture occurred. The fractures occurred at several locations in the axial direction simultaneously. This fact shows clearly that the homogenous compaction of the sample powders has been accomplished by our new measuring method proposed here.

In Figure 5, the relations between the plastic deformation coefficient (Y) and porosity are shown for two sample powders at ambient and elevated temperature. Thus, the developed measurement methods of the characteristic rheological parameters Y and σ_f of fine powders at ambient and elevated temperature, which determine the flow characteristics of fine and ultra-fine powders, should be very significant in view of science and in the application of fine and ultra-fine powders. In Figure 6, the relation of the ratio of the compaction stress per unit layer (ΔP^{**}comp.) and the tensile stress generated (σ_f), namely $\sigma_f/\Delta P^{**}$comp. versus porosity of the sample (ϵ) at ambient and elevated temperature for sample powders ($CaSO_4$ and $CaCO_3$) is demonstrated.

CONCLUSION

(1) A novel measurement method to obtain the characteristic rheological parameters (tensile strength σ_f and plastic deformation coefficient Y) of cohesive fine, or/and ultra fine powders at ambient and elevated temperature is summarized here, using anhydrous calcium sulfate and calcium carbonate fine powders as representative sample powders. Because these rheological parameters Y and σ_f can determine the flow characteristics, the method and results of this paper seem to be significant to the science and technology of fine and/or ultra fine powder technology.

(2) Three comprehensive measurement methods of the above rheological parameters are proposed to characterize fine and ultra fine powders.

ACKNOWLEDGMENT

This research was partially supported by U.S. Department of Energy-Morgantown Energy Technology Center, DE-AP21-95MC-10784.

NOTATION

d^*	: effective particle diameter, (m)
H	: height of packed bed powder (m)
k	: Coordination number, (-)
N	: number of unit layer (-)
U	: gas velocity (m/s)
ΔP^*	: gas pressure drop across the bed, (Pa)
ΔP^{**}	: gas pressure drop across the unit layer, (Pa)
Y	: plastic deformation coefficient of filter cakes, (Pa)

Subscript
c	: at compaction
f	: at fracture
lp	: at loosely packed bed

Greek letters
ϵ	: porosity, (-)
μ	: viscosity of the gas, (Pa.s)
ρ	: density of gas, (kg/m^3)
ϕ	: shape factor of the particle, (-)
σ_f	: fracture strength of the filter cakes, (Pa)

LITERATURE CITED

1. Kono, H.O., L. Richman, and E. Koresawa, "Characteristic Criterion of Rheological Parameters to Predict Flow Properties of Fine Powders," AIChE Symp. Ser.. No. 313, vol. 92 p 114-119(1996).
2. Kono, H.O., Y. Itani, E. Aksoy, E. Koresawa, and J.J. Su, "Characterization of Fluidization Properties of Fine Powder FCC Catalyst at Elevated Temperature," AIChE Symp. Ser.., Fluidization and Fluid- Particle Systems No. 308, Vol. 91, p.170-179 (1995).
3. Kono, H.O., E. Aksoy, and Y. Itani, "Measurement and Applications of the Rheological Parameters of Aerated Fine Powders -- A Novel Characterization Approach to Powder Flow Properties," Powder Technology, 81, 177-187 (1994).
4. Kono, H.O., et al., "Quantitative Criteria for Emulsion Phase Characteristics and for the Transition between Particulate and Bubbling Fluidization," Powder Technology, 52, 69-72 (1987).
5. Rumpf. H., "Grundlagen and Methoden des Granulierens," Chem. Ing. Tech., 30, 144 (1958).
6. Smith, D.H., G. Haddad, H.O. Kono, L. Richman,"Progress in Pressurized Fluidized Bed Combustion (PFBC) Filter Cakes and Filter Cleaning," Pittsburgh Energy Conference, (1996).
7. Yokoyama, T., K. Fuji, and T. Yokoyama, Measurement of the Tensile Strength of a Powder Bed," Powder Technology., 32, 55 (1982).

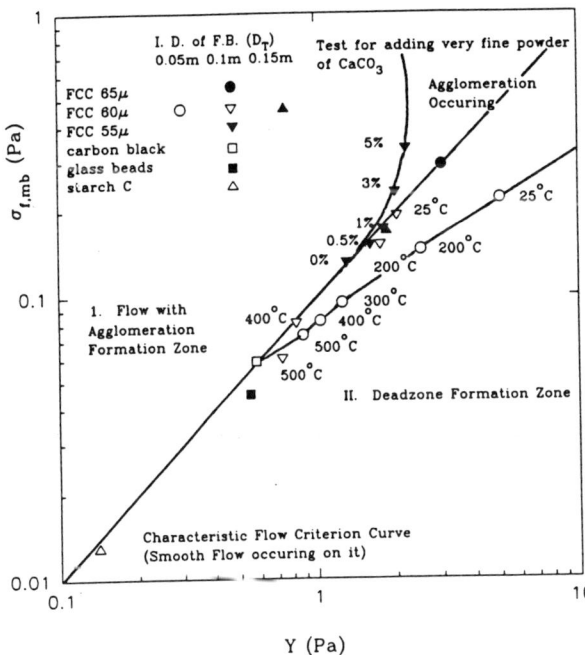

Figure 1: Flow Pattern Diagram of Homogeneously Expanded Fine Dust Powders in Motion with its Characteristic Flow Criterion Curve, Flow with Agglomeration Formation and Deadzone Formation Zones.

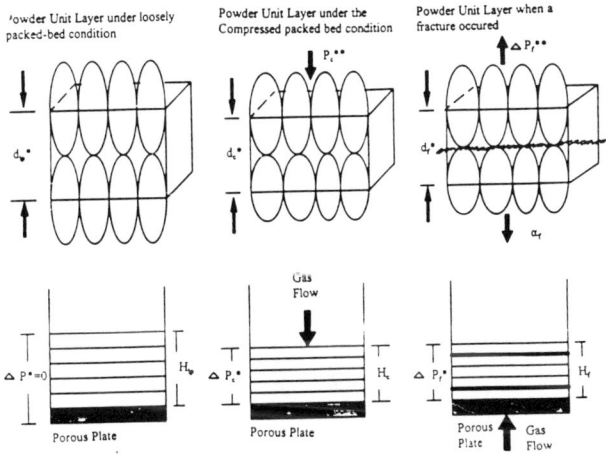

Figure 2: Schematic modeling pictures of the unit layer structures and the beds with the heights of $H_{lp}(d_{lp}^*)$, H_c (d_c^*), and H_f (d_f^*).

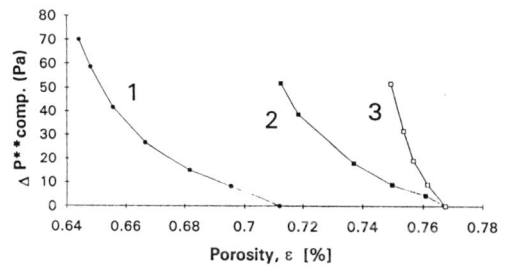

Figure 3: Compression stress per unit layer (ΔP^{**}comp.) versus porosity (ϵ) for (1) calcium carbonate at 25°C, (2) calcium sulfate at 25°C, and (3) calcium sulfate at 542°C.

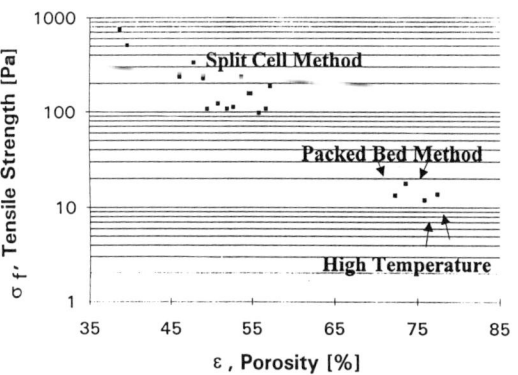

Figure 4a: Tensile strength (σ_f) versus porosity (ϵ) for anhydrous calcium sulfate. High Temperature (542°C) experiment indicated by arrow. All other experiments conducted at 25°C.

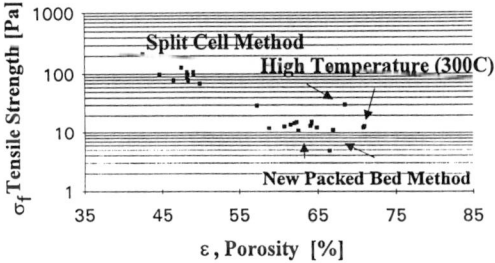

Figure 4b: Tensile strength (σ_f) versus porosity (ϵ) for calcium carbonate. High Temperature (300°C) experiment indicated by arrow. All other experiments conducted at 25°C.

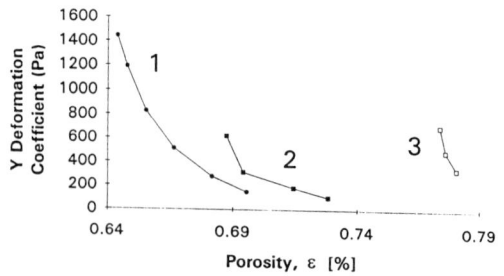

Figure 5: Plastic deformation coefficient (Y) versus porosity (ϵ) for (1) calcium carbonate at 25°C, (2) calcium sulfate at 25°C, and (3) calcium sulfate at 542°C.

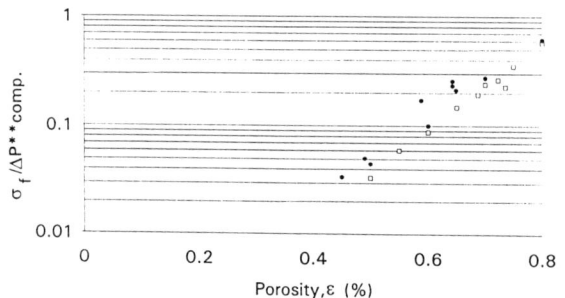

Figure 6: Ratio of the compaction stress per unit layer and the tensile strength ($\sigma_f/\Delta P^{**}$comp.) versus porosity (ϵ) for (●) calcium carbonate at 25°C and (□) calcium sulfate at 25°C.

Local Bubble Properties in 40 and 80 cm Diameter Fluidized Beds

J.S. Groen, J.J.L. Berben, R.F. Mudde and H.E.A. Van Den Akker
Kramers Laboratorium voor Fysische Technologie
Delft University of Technology Prins Bernhardlaan 6, NL-2628 BW Delft, The Netherlands

In this paper we report on experimental results of local bubble properties, measured with a fibre optic probe in gas-solid fluidized beds of 40 and 80 cm diameter. Results are presented on bubble fraction, bubble velocity and bubble chord length. From the bubble chord length distribution, the bubble size distribution can be determined. From correlation functions of the signals of either one or two probes it is possible to determine characteristics of the bubble flow in the fluidized beds. The results lead to a general view of the hydrodynamic behavior of gas-solids fluidized beds.

INTRODUCTION

Fluidized beds are used extensively in process industry as gas-solid, liquid-solid or gas-liquid-solid contactors. Their major advantages are the absence of moving parts and their excellent mixing properties. The mixing behaviour is believed to mainly result from the stirring effects of bubbles (see *e.g.* Kunii and Levenspiel [1]). These occur in fluidized beds at a certain value of the superficial gas velocity U (in excess of the minimum fluidization velocity u_{mf} of the powder). In order to fully understand the hydrodynamic behaviour of a bubbling fluidized bed (or, for that matter, of *any* two-phase reactor), it is of paramount importance to know local and dynamic values of its hydrodynamic parameters. In the case of a gas-solid fluidized bed, important parameters are those describing the bubble dynamics. In an earlier paper (Groen *et al.* [2]) we reported on the development and testing of the "Double Horseshoe Probe", an intrusive fibre optic technique capable of measuring local bubble dynamics and with only small disturbance on the flow field.

EXPERIMENTAL

An overview of the technique is shown in Figure 1: Figures 1a and b show the two

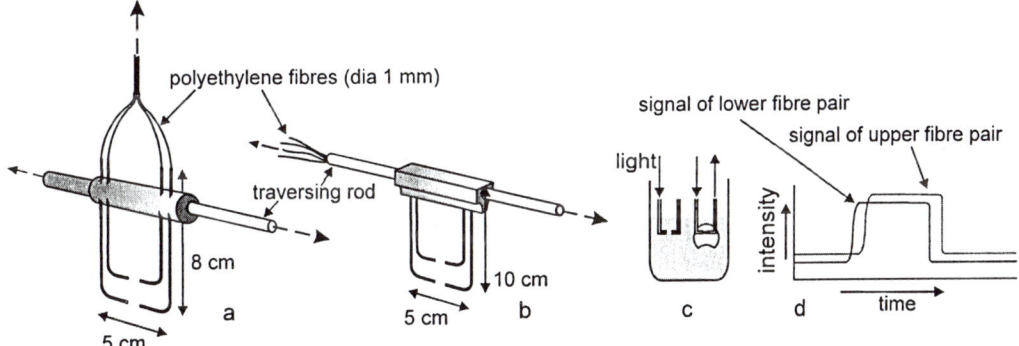

Figure 1: Overview of the measuring technique; a: the double horseshoe probe as used in the 40 cm bed; b: as used in the 80 cm bed; c: principle of operation; d: bubble passing signal.

versions of the probes that were constructed; the fibres used were polyethylene fibres of 1 mm diameter. The fibres were covered with a 0.6 mm protective layer. Figure 1c shows the principle of operation: light is sent into the "emitting" fibre; when powder is present between the leg pair, the "receiving" leg does not detect any light. When a bubble is present at the probe, light is detected by a light sensitive cell at the end of the receiving fibre. Two sets of legs are used within one probe, which by evaluation of the time shift between the signals of the two pairs makes it possible to measure the vertical component of the bubble velocity and hence the bubble chord length. An idealized bubble passing signal is shown in Figure 1d. Two probes were used simultaneously, both were fitted in a traversing system that enabled independent and accurate traversing of the probes accross the column cross sectional area. The horizontal distance between either emitting leg and its corresponding receiving leg was 1 cm, the vertical distance between the two leg pairs was approximately 2 cm.

Experiments were performed in two gas-solid fluidized beds, of 40 and of 80 cm diameter. The 40 cm column was made of perspex, the 80 cm column was made of Corten A steel (Figure 2 shows a picture of the 80 cm column). The 40 cm column was operated at ungassed bed heights of 75 and 175 cm; this height was fixed at 320 cm in the 80 cm column. Both columns were equipped with 5 mm sintered bronze porous

Figure 2: The 80 cm fluidized bed.

plate distributor plates (pore diameter 30 to 70 μm). In both columns, measurements were performed at two heights above the distributor plate: 55 and 155 cm in the 40 cm column and 105 and 255 m in case of the 80 cm column. The powder used was sand of 0.2-0.5 mm diameter and minimum fluidization velocity u_{mf}=9 cm/s (bed density 1584 kg/m^3; Geldart [3] type B powder). The gas superficial velocity U was varied through the bubbling regime, which runs from $U=u_{mf}$ to about $U=2u_{mf}$.

RESULTS

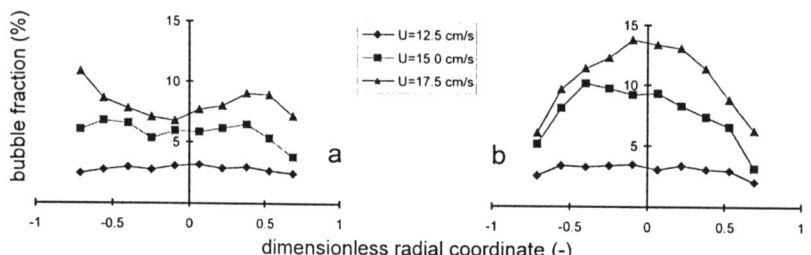

Figure 3: Measured bubble fractions in the 40 cm column. a: 55 cm and b: 155 cm above the distributor plate.

Time averaged results

Figure 3 shows the averaged bubble fraction profiles, as measured at the two levels in the 40 cm column. It can be seen clearly that the profiles become steeper with increasing distance above the distributor plate, in other words a bubble street is forming. Bubble streets cause large-scale overall powder circulation and are in this way important in describing the performance of a fluidized bed reactor. Due to air maldistribution, caused by the porous plate, in the 80 cm column the develop-ment of the bubble fraction profile could not be studied as accurately, though it was observed that in the 80 cm column as well, bubbles tend to travel towards the central region of the bed.

In Figure 4a examples of measured averaged bubble velocity profiles are shown, Figure 4b shows the corresponding bubble velocity distribution. Apart from a prominent peak at low bubble velocities, a tail into higher velocities shows. These high velocities are caused by coalescing bubbles, that, just before coalescing, are accelerated into the wake of a preceding bubble and move very fast. Studying the time signals revealed that practically all fast bubbles followed closely upon another (slower) bubble, so they were indeed in the process of coalescence. With increasing superficial gas velocity the contribution of the lower velocities decreases and the amount of coalescence-related (high) velocities increase. The position of the low-velocity peak changes with increasing superficial gas velocity. This is related to the increase of the bubble size with increasing superficial gas velocity (see Figure 5).

Combining the determined bubble velocities with the bubble passing times one can

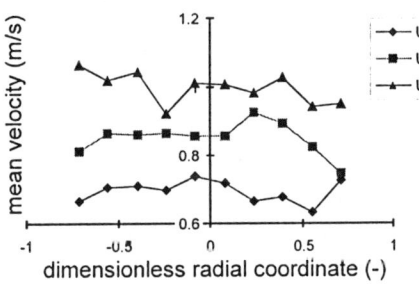 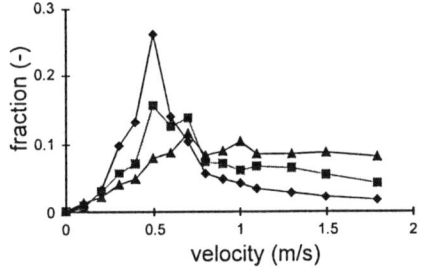

Figure 4: a: Measured bubble velocity profiles; b: corresponding bubble velocity distributions. 40 cm column, 155 cm above the distributor plate.

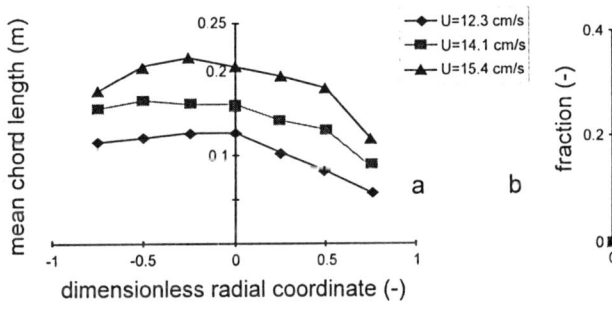 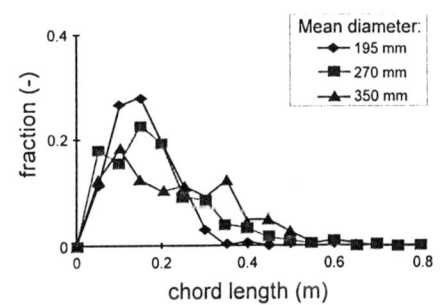

Figure 5: a: Measured mean bubble chord length profiles; b: corresponding chord length distributions in the centre. 80 cm column, 255 cm above the distributor plate.

determine the bubble chord length distribution. Assuming that the bubbles have a fixed shape (but do not necessarily all have the same size) it is possible to calculate the bubble *size* distribution from the bubble chord length distribution (see Clark and Turton [4]). In Figure 5a an example of averaged bubble chord length profiles in the 80 cm column is given, Figure 5b shows the corresponding chord length distributions measured in the centre of the column. From both Figures 5a and 5b it follows that with increasing superficial gas velocity the averaged chord length (and hence the bubble size) increases. The bubble size increases with increasing distance to the distributor plate as well. This is mainly caused by bubble coalescence (see Darton et al. [5]) and is reflected by a decrease of the bubble passing frequency with increasing distance from the distributor plate.

Correlation results
The bubble flow can be studied in a different way by calculating the autocorrelation function of the signal of one pair of legs of a probe. Two examples are shown in Figure 6. Figure 6a shows a case where significant correlation exists only over a small time shift, *i.e.* while the probe is still within the same bubble. This inidicates that the bubble flow is highly irregular. Figure 6b shows an example where a clear periodicity can be seen, or, stated differently, where the bubble flow is more regular. Generally, with increasing distance to the distributor plate the flow becomes more regular. From simple modelling it follows that the characteristic points of the autocorrelation function can be explained in terms of bubble passing time scales: the first zero crossing is related to the averaged bubble passing time and the position of the first maximum refers to the prevailing time between the arrival of two consecutive bubbles. The period of negative correlation directly following the first zero crossing essentially means that it is not likely that two bubbles follow upon each other at a very short distance. The position of this minimum is thus related to the maximum bubble size. Comparing the bubble sizes calculated by means of these time scales to those calculated directly shows that the former gives slightly larger values.

Figure 7 shows the cross correlation functions of the simultaneously measured signals of two probes a horizontal distance apart, at the two distances from the

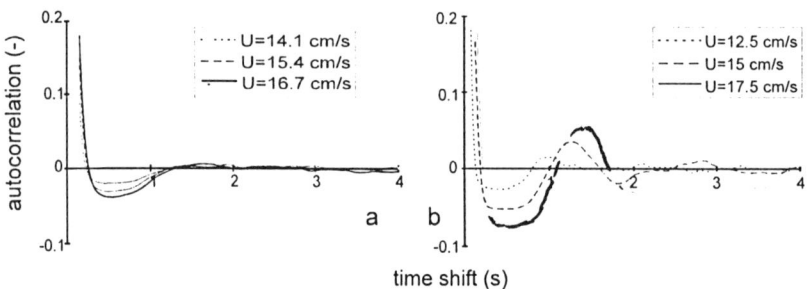

Figure 6: Examples of autocorrelation functions of the signal of a probe; a: irregular flow (80 cm column, 105 cm above distributor, central region); b: regular flow (close to the wall in the 40 cm column, 155 cm above the distributor plate).

distributor plate in the 80 cm column. It is seen that the correlation increases with increasing distance to the distributor. The

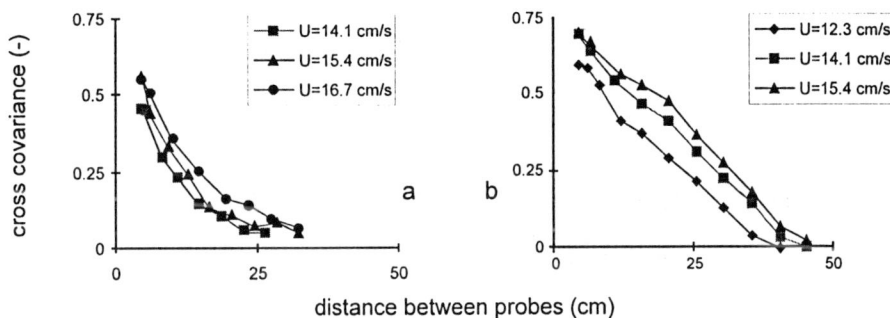

Figure 7: Spatial cross covariance function of the signals of two probes in the 80 cm column, a: 105 cm; b: 255 cm above ths distributor plate.

same effect was observed in the 40 cm column. Close to the distributor plate, the distance over which cross correlation exists is somewhat larger than the maximum size of the bubbles present at the same position, indicating the presence of not only single bubbles. Higher in the column, the distance over which correlation exists only slightly exceeds the bubble size. In this case it can be concluded that single bubbles determine the typical scales of the flow within the fluidized beds and that the occurrence of bubble swarms such as appearing in bubble columns (see Groen et al. [6]) is not likely.

CONCLUSIONS

The Double Horseshoe Probe has proved to be a valuable instrument for measuring local bubble dynamics in fluidized beds. Bubble fraction, bubble velocity and bubble chord length can be accurately measured and effects such as bubble coalescence can be studied. The formation of a bubble street is observed. With increasing superficial gas velocity the bubble size and bubble velocity increase, predominantly caused by an increase in bubble coalescence. From correlations it follows that with increasing distance to the distributor plate the bubble flow tends to become more regular and that single bubbles dominate the flow.

NOTATION

U superficial gas velocity (m/s)
u_{mf} minimum fluidization velocity (m/s)

LITERATURE CITED

[1] Kunii, D. and O. Levenspiel, Fluidization Engineering,, 2nd Ed., Butterworth-Heinemann, Boston (1991).
[2] Groen, J.S., D.C. De Haseth, R.F. Mudde and H.E.A. Van Den Akker, Measurement of Local Fluid Dynamics in 2D- and 3-Bubbling Fluidized Beds, *Proc. 2nd Int. Conf. Exp. Fluid Mech.*, Torino, Italy, July 4-8, 134-141 (1994).
[3] Geldart, D. Types of Gas Fluidization, *Powder Technol.*, **7**, 285-292 (1973).
[4] Clark, N.N. and R. Turton, Chord Length Distributions Related to Bubble Size Distributions in Multiphase Flow, *Int. J. Multiphase Flow*, **4**, 413-422 (1988).
[5] Darton, R.C., R.D. LaNauze and J.F. Davidson, Bubble growth due to coalescence in fluidized beds, *Trans. I. Chem. Engrs*, **55**, 274-280 (1977).
[6] Groen, J.S. R.G.C. Oldeman, R.F. Mudde and H.E.A. Van Dan Akker, Coherent Structures and Axial Dispersion in Bubble Column Reactors, *Chem. Engng Sci.*, **51**(10), 2511-2520 (1996).

A Numerical Model For Wave-Like Gas-Solids Flow In Pipes

David J. Mason and Shi Liu
Centre for Industrial Bulk Solids Handling, Department of Physical Sciences
Glasgow Caledonian University, Cowcaddens Road, Glasgow G4 OBA, Scotland

This paper presents a new model for wave-like gas-solids flow in pneumatic conveying systems. This mode of non-suspension flow can be achieved by many granular materials, such as plastic pellets, and offers the potential benefits of lower rates of particle attrition, or pipeline wear; and in many cases greater energy efficiency. The one-dimensional conservation equations for mass and momentum for the gas and solids phases are solved to determine the variation of flow parameters inside a wave. By predicting the separation of the waves the model can be used to compute the number of waves in a pipe and consequently the flow conditions in the pipe. The prediction of wave velocity and wave distribution show good agreement with experimental observations.

The mode of gas-solids transport in a pneumatic conveying system may be classified as either suspension, or non-suspension flow. Two distinct modes of non-suspension flow exist: moving-bed flow; and wave-like flow. The benefits of transporting a bulk material in a non-suspension flow are a lower rates of particle attrition, or pipeline wear; and in many cases greater energy efficiency.

This paper presents a numerical model for the wave-like mode flow. Materials capable of this mode of flow have been identified in [1] as exhibiting very high permeability. Examples include polyethylene pellets and mustard seeds. Figure 1 illustrates the features of such a flow.

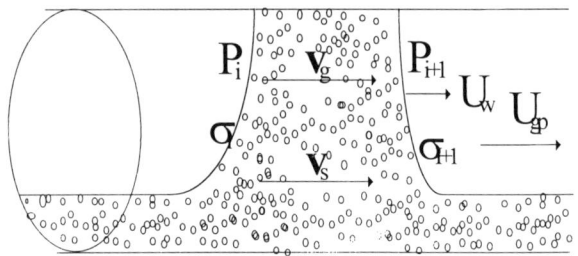

Figure 1   Wave-like gas-solids flow.

Glasgow Caledonian University, Glasgow, Scotland, U.K.

Several workers [2, 3, 4] have described this mode of flow and the shape of the waves. Between each wave is a stationary layer of material. Material is picked up from the stationary layer, transported along the pipe in the wave, and then drops off the back of the wave. The depth of the stationary layer decreases as the flow velocity increases.

THE MODEL

Many of the models used to simulate this type of flow adopt a similar analogy to that used in [5]. These models have several features in common. Firstly, uniform flow properties are assumed throughout a solids wave, thus the Ergun equation can be used to calculate the pressure drop. This could lead to errors if flow properties, such as gas velocity and solids velocity, vary inside a wave. Secondly, in order to obtain a solution the solids velocity must be determined. This is dependent upon the ratio of the area occupied by the stagnant solids layer between waves to the pipe cross sectional area, α. This crucial parameter is, however, based on an analogy to gas-liquid flow. Its formula

$$\alpha = 1 / \left[1 + v_s / \left(0.542\sqrt{gD}\right)\right] \qquad (1)$$

reveals no dependency on the properties of the bulk material. Thirdly, to compute the pressure drop along a pipe, both the length of a single wave and the number of waves in the pipe need to be provided. However the

152

number of waves in the pipe at a particular time is not readily determined. Furthermore, the theory will result in constant gaps between the moving waves, which contradicts the experimental evidence of a varying gap length between waves [6].

The model proposed in this paper takes a fundamentally different approach. Firstly, the flow properties in a solids wave, namely gas and solids velocities, voidage, and pressure, are obtained from numerical solutions of conservation equations. This removes the reliance on an analogy to gas-liquid flow for the area ratio and reflects the variations of the flow parameters inside a wave. Therefore it may result in higher accuracy than the previous models. The conservation equations include:

Continuity

Solids
$$\frac{d(\varepsilon_s v_s)}{dx} = 0 \qquad (2)$$

Gas
$$\frac{d(\rho_g \varepsilon v_g)}{dx} = 0 \qquad (3)$$

Momentum conservation

Solids
$$\frac{d(\rho_s \varepsilon_s v_s^2)}{dx} = -\frac{d\sigma}{dx} - \varepsilon_s \frac{dP}{dx} + \beta_a (v_g - v_s) - \frac{4\tau_{sw}}{D} \qquad (4)$$

$$\frac{d\sigma}{dx} = -G \frac{d\varepsilon_s}{dx} \qquad (5)$$

$$G = 10^{8.76\varepsilon_s - 3.33} \qquad (6)$$

$$\tau_{sw} = \mu_w k_w \sigma + \tan\phi_w (1 + k_w) c \cos(\omega + \phi_w) + \tan\phi_w \frac{D}{2} \varepsilon_s \rho_s g + c_w \qquad (7)$$

$$\beta_a = 150 \frac{\mu \varepsilon_s^2}{\varepsilon d^2} + 1.75 \frac{\rho_g \varepsilon_s |v_g - v_s|}{d} \qquad (8)$$

Gas
$$\frac{d(\rho_g \varepsilon v_g v_g)}{dx} = -\varepsilon \frac{dP}{dx} - \beta_a (v_g - v_s) - \frac{4\tau_{gw}}{D} \qquad (9)$$

$$\tau_{gw} = \frac{f_g}{2} (\rho_g \varepsilon) v_g^2 \qquad (10)$$

$$f_g = 16/\text{Re} \qquad \text{for Re} < 2100$$
$$f_g = 0.0791/\text{Re}^{0.25} \qquad \text{for Re} \geq 2100 \qquad (11)$$

The model assumes that:

- the flow is one-dimensional and steady;
- the bulk material is cohesionless, which is the case for materials capable of this mode of flow;
- the flow is in a horizontal pipe;
- the temperature is constant;
- the fluid is treated as perfect gas.

Examining the momentum conservation equation for the solids it can be seen that descriptions for three phenomena are required:

- The normal stress within the wave of solids. This was determined using the bulk modulus for the powder, G, which relates the normal stress, σ, to the solids concentration, ε_s.

- The particle-wall friction, which was described by a combination of Coulomb friction and particle wall cohesion.

- The inter-phase momentum transfer. The inter-phase friction coefficient, β_a, used to describe this phenomena was derived from the Ergun equation.

For the gas phase, wall friction is calculated according to Darcy's law.

The following boundary conditions were used:

- at the inlet, values are specified for pressure, temperature, voidage, gas and solids mass flow rate;
- at the outlet, pressure and gas velocity are allowed to float.

The normal stress is determined according to [5]:
$$\sigma = \varepsilon_s \rho_s \alpha U_w v_s \qquad (13)$$

which is derived from the momentum balance at the front of a wave. Values of other variables are obtained by zero, or first order extrapolations.

Existing models have used an analogy to gas-liquid flow to determine the area ratio in equation 13. In this work a new approach was used to determine α. The relationship between the normal stress and the velocity of a stress wave inside the solids, $\sigma = v_s\sqrt{G\rho_s}$ [7], was combined with Equation 13 in order to derive a relationship for U_w, v_s and α:

$$U_w = v_s / (1-\alpha) \quad (14)$$

manipulating the above equations yields the equation for area ratio:

$$\alpha = 1/(1 + v_s / v_c) \quad (15)$$

where v_c is the velocity of a stress wave inside the solids wave:

$$v_c = \sqrt{G}/\sqrt{\varepsilon_s \rho_s} \quad (16)$$

The equation takes the same form as Equation 1, but includes the properties of the bulk material. It agrees with the experimental discovery that the stagnant layer diminishes as solids velocity increases.

The conservation equations were solved using a Pressure Correction scheme SIMPLEC [8].

SINGLE-WAVE

Figure 2 shows the prediction of the variation of flow conditions through two adjacent waves 1 m long in a pipe with D = 0.081m.

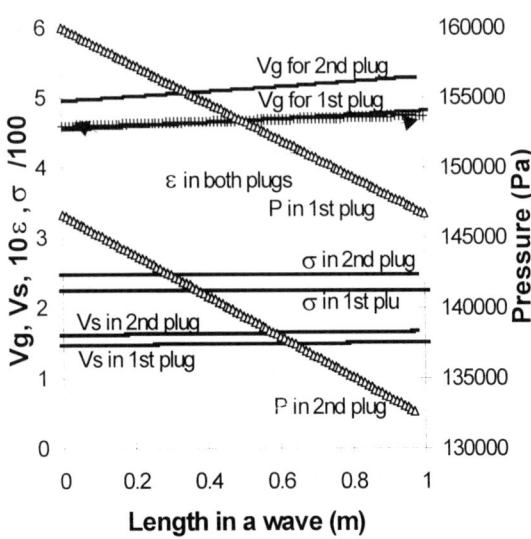

Figure 2 v_g, v_s, ε, σ and P in two adjacent waves

The inlet pressure is 1.6 bar$_a$, and the mass flow rates of the solids and gas are 0.3 kg/s and 0.02 kg/s respectively. The bulk material simulated was plastic pellets with d = 3mm, ρ_s = 880kg/m^3, ϕ = 45° and ϕ_w = 15°. The voidage, ε, at the start of each wave is 0.45.

The solids velocity and normal stress profiles undergo a slight increase, but the gas velocity increases noticeably through a wave. This implies a varying slip velocity between the gas and the solids through a wave. The flow conditions in the two waves are markedly different: both solids and gas velocities, as well as normal stress are higher in the second wave. This indicates a possible source of error in previous models that treat the flow conditions as uniform throughout all the solids waves.

PIPE SECTION

The single wave calculation can be extended to predict the pressure drop along a pipe containing several waves. The inlet conditions to the pipe section are used to determine the inlet boundary conditions for the first wave in the pipe. The length of the gap to the next downstream wave may be determined from the ratio of the average solids mass flow rate (specified) to the solids mass flow rate in the wave:

$$\beta = \frac{\dot{m}_s}{A(1-\varepsilon)\rho_s v_s} \quad (17)$$

The inflow pressure for the second wave can then be determined. This process is repeated until the outlet of the pipe section is reached.

Figure 3 shows some typical results for an 80 m pipe. It shows the variations of gas and solids velocities within each solids wave, an increasing pressure drop across each wave, and the expansion of the gaps between the waves down stream. Such results agree with experimental observations, unlike the predictions of previous models

CONCLUSIONS

A new one-dimensional model for wave-like gas-solids flow in pneumatic conveying systems has been presented. This model allows the velocities associated with the wave (gas, solids, wave front) to be evaluated without recourse to an analogy with gas-liquid flow. It predicts a variation of flow parameters inside a wave, as well as an increasing gap length between waves, which agrees with experimental discoveries. This model can also predict the number of waves in a pipe and consequently the flow conditions in the pipe, while earlier models require this as an input parameter.

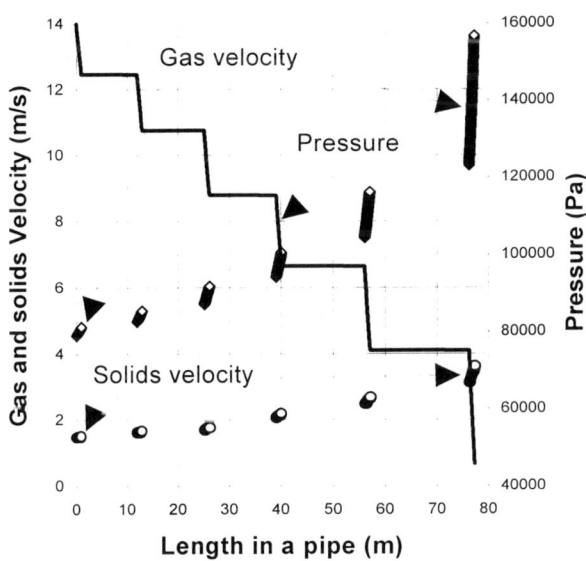

Figure 3 v_g, v_s, ε, σ and P in a 80 meter pipe.

NOTATION

A	pipe cross sectional area (m^2)	U_w	wave velocity (m/s)
c	inter particle cohesion (Pa)	v	velocity (m/s)
c_w	particle-wall cohesion (Pa)	x	co-ordinate (m)
D	pipe internal diameter (m)	α	area ratio (-)
d	particle diameter (m)	β	length ratio (-)
f	Darcy friction factor (-)	ε_s	solids voidage (-)
G	compressibility modulus for solids (Pa)	ε	voidage $\varepsilon = 1 - \varepsilon_s$
g	gravitational acceleration (m/s^2)	τ	shear stress (Pa)
k	coefficient of internal friction	σ	normal stress (Pa)
k_w	K at wall	ϕ	internal friction angle (degree)
$\dot{m}$	mass flow rate (kg/s)	ϕ_w	wall friction angle (degree)
P	pressure (Pa)	μ	gas viscosity (Pa s)
Re	pipe Reynolds number	ρ	density (kg/m^3)

Subscripts
g: gas; s: solids; w: wall ; gw: gas phase at wall, sw: solids phase at wall

LITURATURE CITED

1. Jones, M.G., and Mills, D., 1990, "Product Classification for Pneumatic Conveying," *Powder Handling and Processing*, Vol. 2, No. 2, pp. 117-122.

2. Konrad, K., 1988, "Boundary Element Prediction of the Free Surface Shape between Two Particle Plugs in a Horizontal Pneumatic Transport Pipeline," *Canadian Journal of Chemical Engineering*, Vol. 66, pp. 177-1812.

3. Madhusudana Rao, M., Ramakrishnan, T., and David, P., 1991 "Numerical Study of Plug-Type Pneumatic Conveying," *Powder Handling and Processing*, Vol. 3, No. 1, pp. 57-60.

4. Tsuji, Y., Tanaka, T., and Ishida, T., 1992, "Lagrangian Numerical Simulation of Plug Flow of Cohesionless Particles in a Horizontal Pipe," *Powder Technology*, Vol. 71, pp. 239-250.

5. Konrad, K., Harrison, D., Nedderman, R.M, and Davidson, J.F., 1980, "Prediction of the Pressure Drop for Horizontal Dense Phase Pneumatic Conveying of Particles" *Pneumotransport 5* pp225-244.

6. Mason, D.J., Cairns, C., Knight, E.A., and Pugh, J.R., 1995 "An Investigation of Wave-Like Gas-Solids Flow in Pipelines Using a Variety of Measurement Techniques" *ASME Heat Transfer and Fluids Engineering Divisions*, HTD-Vol. 321/FED-Vol. 233, pp. 525-29.

7. Das, Braaja M. 1993, "Principles of Soil Dynamics", PWS-KENT Publishing Company, Boston, USA, ISBN 0-534-93129-4.

8. Patankar, S.V., 1980, "Numerical heat transfer and fluid flow", Hemisphere Publishing Corporation, Taylor & Francis Group, New York.

Development of a Solids Flowmeter for an Industrial Scale Operation

John Kost and Jack Saluja
Viking Systems International, 2070 William Pitt Way, Pittsburgh, PA 15238

Jonathan Thorn, William Link and George Klinzing
Chemical and Petroleum Engineering Department, University of Pittsburgh, Pittsburgh, PA 15260

There have been numerous attempts to develop a simple solids flowmeter that could respond adequately to the industrial environment of pneumatic conveying of solids. A reliable and economical commercial solids flowmeter for measuring solids flow in pneumatic systems has been developed based on pressure drop experienced in the flow of the gas-solid suspension. The flowmeter is on-line, non-intrusive, accurate, inexpensive and reliable. In 1992, Cabrejos and Klinzing (1992) first tested the concept of a flow meter that was based on the pressure loss across a straight section of piping when a gas-solids suspension was flowing. These initial tests were carried out for different materials for a 2 inch diameter conveying system. These researchers showed that the response's specific pressure drop with the solids loading was linear for the several materials tested. They also developed the basic analysis for this phenomenon. In order to expand the concept to an industrial application, large diameter flowmeters were needed. Experiments have been conducted using three different flow loops testing different solid materials. Two test loops tested provided up to 1,800 lbs/hr of material in a four inch diameter system while a larger commercial scale unit tested up to 15,000 lbs/hr in a 4 inch diameter unit. The larger facility tests were kindly provided by DuPont. A linear correlation of the response of the measured and predicted feed rate were achieved in both test series. The results obtained in this study are encouraging for the industrial application of the flowmeter.

The measurement of solids flow in pneumatic conveying is an elusive characteristic that is very much more complex to address than the measurement of the flow of a gas and liquid stream by themselves. The varying properties of the solids cause variations from measurement system to system. Over the years a number of devices have been proposed but none have surfaced as an instrument that industry has enough confidence in to set the standard for solids flow rates. One of the biggest problems in devices is the degree of calibration that needs to be performed in order to employ the instrument. We have developed a simple flow meter for the dilute phase flow of solids in a gas stream that is based on the fundamental physics of the flow process. The pressure drop over a standard section of piping can be used as the basis for a solids flow meter. Recent tests at DuPont have shown that the new device is able to measure the flow of particles at a rate of 15,000 lbs/hr in the dilute phase condition with good accuracy and reproducibility.

In all of solids processing operations, the flow rate of solids is an important parameter that will give the operator the information that is needed to control and operate various section of the plant that have solids present. Presently, the generally accepted practice is to employ weigh cells between the two units such that changes in the weight over a fixed period of time can give one the solids flow rate. This type of measurement is far from instantaneous and is not suitable for operating a control of the process.

Proposed solid flow meters have come in a wide assortment of concepts based on mechanics, light, sound, electrostatics, magnetic, field forces and combinations of these phenomenon. There is a tremendous need to have a device to measure the solids flow rate. Some unique characteristics that would be desirable in such a meter would be:

- High accuracy and reproducibility
- Rugged
- Little or no calibration needed
- Low cost
- Functions over a wide temperature range
- Non intrusive in nature
- Independent of gas and solids properties

In 1924 a set of experiments by Gasterstadt (1924) showed that if one plotted the relative pressure drop versus the solids loading for a lightly loaded condition, a straight line presentation could be obtained.

More recently Cabrejos and Klinzing (1992) applied this condition to a wide variety of data including some testing of their own. Figure 1 shows an example of one of these series of tests which was analyzed using glass beads. As one can see, the linearity seen is pronounced for the dilute phase regime.

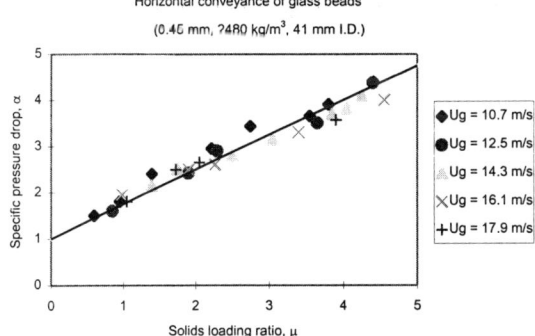

Figure 1. Data from Cabrejos and Klinzing (1992).

The basic physics behind this relationship is a simple pressure drop analysis. For dilute flow one can assume a linear combination of pressure losses due to the gas and solids individually thus:

$$\frac{\Delta P_t}{L} = \frac{\Delta P_s}{L} + \frac{\Delta P_g}{L} \quad (1)$$

This equation can be rearranged to place it in a specific pressure drop format yielding:

$$\alpha = \frac{\Delta P_t/L}{\Delta P_g/L} = 1 + \frac{\Delta P_s/L}{\Delta P_g/L} \quad (2)$$

The pressure losses due to the friction of the gas and solid stream can be related to the diameter of the pipe, D, the length of the pipe, L, the average velocity, U_g, the voidage of the flow, ε, and the friction factors for each phase, f_g and f_s. Using the definition of the voidage for a uniform flow as:

$$\varepsilon = 1 - \frac{4W_s}{\pi D^2 \rho U_p} \quad (3)$$

Replacing the expressions for the frictional terms and voidage in equation 2 gives:

$$\alpha = 1 + \frac{f_s}{f_g} \frac{U_p}{U_g} \mu \quad (4)$$

where μ is the solids loading. In this expression use of the fact that $\rho_s U_p \gg \rho_g U_g$ is employed. Using a nondimensional format one can write:

$$\alpha = 1 + (\frac{f_s}{f_g})(\frac{Re_p}{Re})(\frac{D}{d_p})\mu \quad (5)$$

EXPERIMENTAL

Experiments to test the flowmeter were conducted on three different systems, two at the University of Pittsburgh and one at the DuPont Experimental Station. The first pneumatic conveying system utilized for testing was constructed of four inch copper pipe and was originally designed for use with coarse particles. The system includes a blower capable of 325 scfm of air, a rotary airlock feeder capable of up to 1,800 lbs/hr, and a cyclone for air cleaning. The pipeline configuration is 43 feet horizontal into an 18 foot vertical section followed by another 43 feet horizontal. The particles used in this loop are polyethylene pellets with an average size of 90 percent greater than 2.4 mm.

The second system at the University of Pittsburgh was located in the same lab as the first system and shared the blower. The second system was designed for smaller particles and was originally a two inch system. Almost all of the two inch pipe was replaced with 4 inch copper pipe. The configuration of the system is a 20 foot horizontal section into 18 feet of vertical followed by a 100 foot horizontal section with three bends. The system was equipped with a screw feeder capable of handling up to 1,080 lbs/hr of polystyrene particles, 80 percent of which is greater than 297 microns. Finally, the system finished with a cyclone designed for handling smaller particles.

The DuPont system is a four inch schedule 40 Aluminum pipe system. This system is equipped with a bag house filter and rotary airlock to recycle the particles to a screw feeder used to inject the particles into the conveying. The air mover is blower capable of moving 600 scfm of air. The conveying line configuration is a 32 foot horizontal section followed by an 18 foot vertical section, ending with a 28 foot horizontal section. Two different screw feeders were used which allowed a maximum capacity of 15,000 lbs/hr of Zytel pellets measuring approximately 4mm square by 2mm thick.

RESULTS

Results from the three systems show good linear correlations between the solids loading ratio and specific pressure and can be seen in Figures 2-4. Each of the points on the graphs from the University of Pittsburgh represent an average of 60 points over an hour period and data from the DuPont system are averages of 20 points. The grouping of points on the DuPont graph between 0-3 and 4-6 loadings represent data from two different screw feeders.

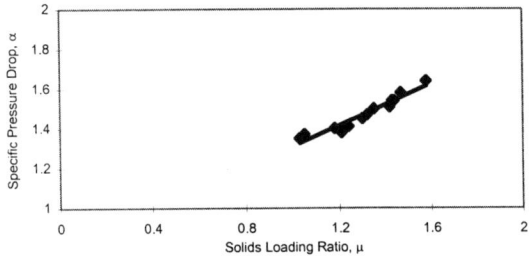

Figure 2. Flowmeter diagram from polyethylene tests at the University of Pittsburgh

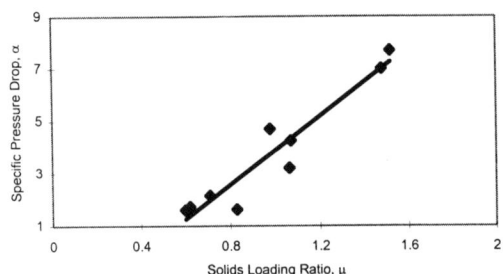

Figure 3. Flowmeter diagram from polystyrene tests at the University of Pittsburgh

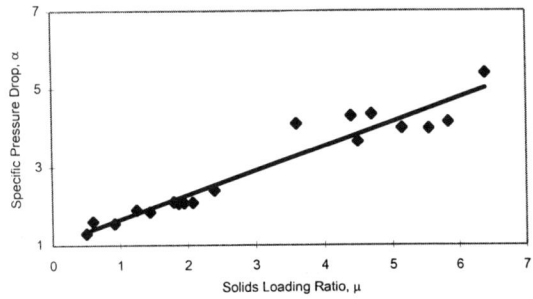

Figure 4. Flowmeter diagram from tests at DuPont experimental station using Zytel pellets

CONCLUSIONS

Data has been taken on two academic systems and one commercial sized system to show that the flowmeter works well for dilute phase pneumatic transport of non-fine (over 100 micron) particles. Three different sizes of plastic particles were tested; all with good linear correlations. The results from this paper show that the flowmeter presents an accurate, reliable, inexpensive solution to solids flow measurement heretofore virtually unknown in the field.

NOMENCLATURE

D - Pipe diameter
d_p - Particle diameter
f_g - Gas friction factor
f_s - Solids friction factor
L - Pipe length
P - Pressure
U_g - Gas velocity
U_s - Solid velocity
W_s - Solids flow rate

Greek

α - Specific pressure drop ratio
ε - voidage

LITERATURE

1. Cabrejos, F.J., Klinzing, G.E. ASME 92-WA/FE-3 (1992).

2. Gasterstadt, J., V.D.I. Zeitschrift, Vol 68, No. 24 (1924), pp. 617-624.

3. Rizk, F., Doktor-Ingenieurs Diss., University of Karlsruhe, (1973).

ACKNOWLEDGEMENTS

The authors would like to acknowledge Timothy Bell and DuPont at the DuPont Experimental Station in Wilmington, Delaware the use of their test facility and helpful insight into the project.

Index

A
Acoustic pulsed waves ... 136
Air classifier, zigzag ... 87
Anemometry, laser doppler (LDA) ... 7
Axial solids distribution ... 97

B
Bottom bed dynamics ... 97
Bubble properties ... 147

C
Capacitance probe measurements ... 55
Circulating fluidized bed[s] (CFB) ... 25, 31, 97, 119
Coating apparatus, Wurster-type ... 125
Cohesive properties of particulate beds ... 92
Combustor application, CFB ... 97

D
Diagnostics, laser doppler anemometry (LDA) ... 7
Down flow reactors ... 46, 61
Draft tube ... 131

E
Elastic properties of particulate beds ... 92
Electrostatic effects ... 119

F
Fines migration in sandstone ... 72
Flow
    behavior ... 25
    gas-solids ... 1
    in the entrance selection of a FCCU ... 36
    properties ... 141
    reactors ... 46
Fluid
    catalytic cracking unit (FCCU) ... 36
    particle interactions ... 103
Fluidized bed(s)
    circulating ... 25, 31, 97, 119
    combustion ... 77
    combustor ... 55, 97
    interconnected ... 40
    local bubble properties in ... 147
    three-phase inverse ... 51
    turbulent, heat transfer in a ... 83
Friction factors in porous media ... 67

G
Gas-liquid
    mass transfer ... 51
    two-phase flow system ... 7
Gas-solids
    flow ... 1
    pipe flow ... 136
    reactors ... 61
    wave-like flow ... 152
Geldart B/A boundary ... 18
Granular
    convection ... 109
    flows ... 113
    temperature, particle ... 18

H
Heat transfer, wall-to-bed ... 83
Hydrodynamic aspects of gas-solids reactors ... 61

I
Interactions, fluid-particle and particle-particle ... 103

K
Kinetic theory ... 12

L
Liquid
    gas, two-phase flow system ... 7
    solid fluidization using kinetic theory ... 12

M
Magnetic resonance imaging ... 109
Mass transfer, gas-liquid ... 51
Metal and sulfur capture by lime ... 77

P
Particle
    circulation ... 131
    clustering ... 46
    granular temperature ... 18
    -particle interactions ... 103
Particulate
    beds, cohesive properties of ... 92
    systems ... 103
Pipes, gas-solids flow in ... 152
Porous media, friction factors in ... 67
Probe measurements, capacitance ... 55

R
Reactors
    down flow ... 46

S
Sandstone, fines migration in ... 72
Scale model, volume average ... 72
Similitude studies, CFB ... 31
Slow shearing granular flows ... 113
Solids
    distribution ... 97
    flowmeter, development of a ... 156
    flux ... 31
    loading ... 31
    residence time distribution ... 40
    volume concentrations and velocities ... 55
Spouted bed, particle circulation in a ... 131
Sulfur and metal capture by lime ... 77

T
Temperature, ambient and elevated ... 141
Tube, draft ... 131

W
Waves, acoustic pulsed ... 136
Wurster-type coating apparatus ... 125

Z
Zigzag air classifier, experimental study of ... 87